James Haddon

Examples and solutions of the differential calculus

James Haddon

Examples and solutions of the differential calculus

ISBN/EAN: 9783742891556

Manufactured in Europe, USA, Canada, Australia, Japa

Cover: Foto ©berggeist007 / pixelio.de

Manufactured and distributed by brebook publishing software
(www.brebook.com)

James Haddon

Examples and solutions of the differential calculus

EXAMPLES AND SOLUTIONS

OF THE

DIFFERENTIAL CALCULUS.

BY

JAMES HADDON, M.A.,

Second Mathematical Master of King's College School.
Author of " Rudimentary Algebra," " Rudimentary Arithmetic," " Rudimentary
Book-keeping," &c &c.

New Edition.

LONDON:

VIRTUE AND CO., 26, IVY LANE,

PATERNOSTER ROW.

INTRODUCTION.

THE Doctrine of Limits is now very generally adopted as the basis of the Differential and Integral Calculus.

Of the methods which were formerly in use it may be advantageous to the mathematical student to glance at some of the most prominent.

By inscribing successively in a circle, regular polygons of four, eight, sixteen, thirty-two, &c. sides, we may at length suppose a polygon to be inscribed whose area shall be less than that of the circle by a quantity so small as to be unassignable. In this manner the area of the circle may be said to be *exhausted.* Hence, the method which was based upon this mode of operation was termed the *Method of Exhaustions.*

In the early part of the seventeenth century a work was published, in which all quantity was assumed to be composed of elements so small that it would be impossible to *divide* them. An infinite number of points in continued contact were supposed to form a line, an infinite number of lines to form a surface, and an infinite number of surfaces to form a solid. Now, since a line has magnitude, namely, length, and a point has no magnitude, it is obvious that a line cannot properly be considered to be made up of a series of points. The method founded upon these suppositions is consequently objectionable. Cavalerius, the inventor of it, called his work " Geometria Indivisibilibus ;" and hence this method was styled the *Method of Indivisibles.*

Sir Isaac Newton considered all quantity to be generated by motion ; a point in motion producing a line, a line in motion producing a surface, and a surface in motion producing a solid. This motion or *flowing* of a point, a line, and a surface, gave rise to the terms *fluents* and *fluxions :* the quantity generated by the motion being called the *fluent* or flowing quantity, and the velocity of the motion, at any instant, the *fluxion* of the quantity generated at that instant. The method founded upon these considerations has been long known as the *Method of Fluxions.*

As applications of this method are continually met with in mathematical works, it may not be inappropriate to give a few instances of its notation, compared with that proposed by Leibnitz, and now generally adopted by writers on the Differential Calculus :

$$\dot{u}, \quad \ddot{u}, \quad \overset{\cdot\cdot\cdot}{u}, \quad \overset{\cdot\cdot\cdot\cdot}{u}, \quad \overset{n}{\dot{u}}, \quad \overline{\sin x}, \quad \left\{ (x^2-1)^m \right\}^{\overset{n}{\cdot}}.$$

$$du, \quad d^2u, \quad d^3u, \quad d^4u, \quad d^n u, \quad d\sin x, \quad d^n (x^2-1)^m.$$

The fluxional symbols in the first line are placed exactly over the corresponding differential symbols in the second.

Leibnitz considered every magnitude to be made up of an infinite number of infinitely small magnitudes. His mode of reasoning was as follows. Any quantity u consists of an infinite number of *differentials*, each equal to $ph + qh^2 + rh^3 + \&c.$, and h being infinitely small, each term in the series is infinitely less than the next preceding term, and consequently the sum of the terms after the first is infinitely less than that first term. Hence ph is the only term necessary to be retained to represent the series.

Lagrange, in his "Calcul des Fonctions," endeavoured to simplify the subject by rejecting the consideration of infinitely small differences and limits, referring the Differential Calculus to a purely algebraic origin. He *defined* the differential of a quantity to be the first term of the series $ph + qh^2 + rh^3 + \&c.$ This is the foundation of his theory.

Each of these methods has found numerous advocates among mathematicians, a fact which excites no surprise when we consider the extraordinary genius of the great men whose names are associated with the origin of these various and most interesting theories.

In our own day several highly talented men have directed their attention to this subject, and it seems now to be very generally admitted that the method best adapted to elementary instruction is that founded on the Doctrine of Limits.

Among the valuable works which have recently enriched this subject may be mentioned those of Whewell, Hall, O'Brien, De Morgan, Thomson, Young, Price, and Walton, in our own language, and Duhamel, Cauchy, Moigno, and Cournot, in the French.

Let us suppose a certain magnitude u to be dependent for its value upon some *variable* magnitude x, so that the value of u may be represented by some expression into which x enters, then u is a *function* of x. We will assume, for instance, that $u = x^3$, and, in this simple example, supposing x to undergo a change of value, we will trace the corresponding effect produced upon the function u.

Let x take the increment h, that is, let x change its value and become $x+h$, then if we represent the corresponding value of u by u_i, we shall have

$$u_i = (x+h)^3 = x^3 + 3x^2h + 3xh^2 + h^3,$$

$\therefore\ u_i - u = 3x^2h + 3xh^2 + h^3 = $ corresponding increment of u,

$\dfrac{u_i - u}{h} = 3x^2 + 3xh + h^3 = $ ratio of increment of function to increment of variable.

Now the first term of this expression for the ratio being $3x^2$, it is obvious that h may undergo any change of value whatever, without affecting this first term.

Let h then continually decrease in value until it is $=0$, then the expression for the ratio will be simply $3x^2$. Hence this first term is the *limit* towards which the ratio approaches as h is diminished, and which limit the ratio cannot reach until $h=0$.

Now if $u = x^3$, $\quad du = 3x^2 \cdot dx$, $\quad \dfrac{du}{dx} = 3x^2$, where du is the *differential* of u, dx the *differential* of x, and $\dfrac{du}{dx}$ the *differential coefficient* derived from the function, that is the coefficient of dx. Thus the *limit* $3x^2$ is equal to the *differential coefficient*.

These remarks are offered to the reader in this place, not only with a view to remind him of what the Method of Limits is, and to regard it in its connexion with the methods above alluded to, but also in the hope of inducing him constantly to recollect that, when he is performing that very common operation in the Differential Calculus of ascertaining the differential coefficient, he is virtually seeking the *limit* of the ratio of the increment of the function to the increment of the variable.

CONTENTS.

CHAPTER I.

DIFFERENTIATION OF FUNCTIONS OF ONE VARIABLE.

Ex. (1.) Let $u = ax$. Then $\dfrac{du}{dx} = a$.

(2.) Let $u = a + 4x$. Then $\dfrac{du}{dx} = 4$.

(3.) Let $y = 3ax^2 + l^2$. Then $\dfrac{dy}{dx} = 2 \times 3ax = 6ax$.

(4.) Let $u = \sqrt{x^3 - a^3}$. Then $\dfrac{du}{dx} = \dfrac{3x^2}{2\sqrt{x^3 - a^3}}$.

(5.) Let $u = \dfrac{2x^4}{a^2 - x^2}$. Then $\dfrac{du}{dx} = \dfrac{(a^2 - x^2).\,8x^3 - 2x^4.\,(-2x)}{(a^2 - x^2)^2}$

$$= \frac{8a^2x^3 - 8x^5 + 4x^5}{(a^2 - x^2)^2} = \frac{8a^2x^3 - 4x^5}{(a^2 - x^2)^2}.$$

(6.) $u = (1 + 2x^2).\,(1 + 4x^3) = 1 + 2x^2 + 4x^3 + 8x^5$.

$\dfrac{du}{dx} = 4x + 12x^2 + 40x^4 = 4x\,(1 + 3x + 10x^3)$.

B

(7.) $u=(1+x)^4.\,(1+x^2)^2.$

$$\frac{du}{dx}=(1+x)^4.\,2(1+x^2).\,2x+(1+x^2)^2.\,4(1+x)^3$$

$$=4(1+x)^3.\,(1+x^2).\,\{(1+x)x+(1+x^2)\}$$

$$=4\,(1+x)^3.\,(1+x^2)\,\{1+x+2x^2\}.$$

(8.) $u=(x^3+a)\,(3x^2+b).$

$$\frac{du}{dx}=(x^3+a).\,6x+(3x^2+b).\,3x^2$$

$$=6x^4+6ax+9x^4+3bx^2=15x^4+3bx^2+6ax.$$

(9.) $u=(a+bx^m)^n.$

$$\frac{du}{dx}=n\,(a+bx^m)^{n-1}.\,mbx^{m-1}=bmn\,(a+bx^m)^{n-1}.\,x^{m-1}.$$

(10.) $u=\left\{a+\sqrt{b+\dfrac{c}{x^2}}\right\}^4$

$$\frac{du}{dx}=4\left\{a+\sqrt{b+\frac{c}{x^2}}\right\}^3.\,\frac{\dfrac{-c.\,2x}{x^4}}{2\sqrt{b+\dfrac{c}{x^2}}}=-\frac{4c\left\{a+\sqrt{b+\dfrac{c}{x^2}}\right\}^3}{x^3\sqrt{b+\dfrac{c}{x^2}}}.$$

(11.) $u=\sqrt{x+\sqrt{1+x^2}}.$ Squaring, we have

$$u^2=x+\sqrt{1+x^2}.$$

$$2u\frac{du}{dx}=1+\frac{2x}{2\sqrt{1+x^2}}=1+\frac{x}{\sqrt{1+x^2}}$$

$$\therefore \frac{du}{dx}=\frac{1+\dfrac{x}{\sqrt{1+x^2}}}{2u}=\frac{1+\dfrac{x}{\sqrt{1+x^2}}}{2\sqrt{x+\sqrt{1+x^2}}}$$

$$=\frac{\sqrt{1+x^2}+x}{2\sqrt{1+x^2}.\,\sqrt{x+\sqrt{1+x^2}}}=\frac{\sqrt{x+\sqrt{1+x^2}}}{2\sqrt{1+x^2}}.$$

$(12.)$ $u=\dfrac{x}{\sqrt{a^2+x^2}-x}$. Multiplying both numerator and

denominator by $\sqrt{a^2+x^2}+x$, we have

$$u=\frac{x\sqrt{a^2+x^2}+x^2}{a^2+x^2-x^2}=\frac{1}{a^2}\left(x^2+x\sqrt{a^2+x^2}\right).$$

$$\frac{du}{dx}=\frac{1}{a^2}\left\{2x+x\cdot\frac{2x}{2\sqrt{a^2+x^2}}+\sqrt{a^2+x^2}\right\}$$

$$=\frac{1}{a^2}\left\{2x+\frac{x^2}{\sqrt{a^2+x^2}}+\sqrt{a^2+x^2}\right\}$$

$$=\frac{1}{a^2}\left\{2x+\frac{x^2+a^2+x^2}{\sqrt{a^2+x^2}}\right\}=\frac{2x}{a^2}+\frac{2x^2+a^2}{a^2\sqrt{x^2+a^2}}.$$

$(13.)$ $u=(a+x)(b+x)(c+x).$

$$\frac{du}{dx}=(b+x)(c+x)\cdot\frac{d(a+x)}{dx}+(c+x)(a+x)\cdot\frac{d(b+x)}{dx}$$

$$+(a+x)(b+x)\cdot\frac{d(c+x)}{dx}$$

$$=(b+x)(c+x)+(c+x)(a+x)+(a+x)(b+x)$$

$$=bc+bx+cx+x^2+ac+ax+cx+x^2+ab+ax+bx+a^2$$

$$=3x^2+2ax+2bx+2cx+ab+ac+bc$$

$$=3x^2+2(a+b+c)x+ab+ac+bc.$$

$(14.)$ $u=(1+x^m)^n.(1+x^n)^m.$

$$\frac{du}{dx}=(1+x^m)^n\cdot\frac{d(1+x^n)^m}{dx}+(1+x^n)^m\cdot\frac{d(1+x^m)^n}{dx}$$

$$=(1+x^m)^n.\,m(1+x^n)^{m-1}.\,nx^{n-1}$$

$$+(1+x^n)^m.\,n(1+x^m)^{n-1}.\,mx^{m-1}$$

$$=mn(1+x^m)^{n-1}.(1+x^n)^{m-1}.\left\{(1+x^m)x^{n-1}+(1+x^n)x^{m-1}\right\}$$

$$=mn(1+x^m)^{n-1}(1+x^n)^{m-1}\left\{x^{m-1}+x^{n-1}+2x^{m+n-1}\right\}.$$

(15.) $u = \sqrt{\left(a - \dfrac{b}{\sqrt{x}} + \sqrt[3]{(c^2 - x^2)^2}\right)^3} = \left\{a - \dfrac{b}{\sqrt{x}} + (c^2 - x^2)^{\frac{2}{3}}\right\}^{\frac{3}{4}}.$

First, $\dfrac{d\left\{a - \dfrac{b}{\sqrt{x}} + (c^2 - x^2)^{\frac{2}{3}}\right\}}{dx} = \dfrac{b \cdot \dfrac{1}{2\sqrt{x}}}{x} + \dfrac{2}{3}(c^2 - x^2)^{-\frac{1}{3}}.(-2x)$

$$= \frac{b}{2x\sqrt{x}} - \frac{4x}{3(c^2 - x^2)^{\frac{1}{3}}}.$$

$\therefore \dfrac{du}{dx} = \dfrac{3}{4}\left\{a - \dfrac{b}{\sqrt{x}} + (c^2 - x^2)^{\frac{2}{3}}\right\}^{-\frac{1}{4}}.\left\{\dfrac{b}{2x\sqrt{x}} - \dfrac{4x}{3(c^2 - x^2)^{\frac{1}{3}}}\right\}$

$$= \frac{3\left(\dfrac{b}{2x\sqrt{x}} - \dfrac{4x}{3\sqrt[3]{c^2 - x^2}}\right)}{4\sqrt[4]{a - \dfrac{b}{\sqrt{x}} + \sqrt[3]{(c^2 - x^2)^2}}} = \frac{\dfrac{3b}{2x\sqrt{x}} - \dfrac{4x}{\sqrt[3]{c^2 - x^2}}}{4\sqrt[4]{a - \dfrac{b}{\sqrt{x}} + \sqrt[3]{(c^2 - x^2)^2}}}.$$

(16.) $u = \dfrac{x}{x + \sqrt{1 - x^2}}.$

$$\frac{du}{dx} = \frac{x + \sqrt{1 - x^2} - x\left(1 - \dfrac{x}{\sqrt{1 - x^2}}\right)}{(x + \sqrt{1 - x^2})^2}$$

$$= \frac{\sqrt{1 - x^2} + \dfrac{x^2}{\sqrt{1 - x^2}}}{2x\sqrt{1 - x^2} + 1} = \frac{1 - x^2 + x^2}{2x(1 - x^2) + \sqrt{1 - x^2}}$$

$$= \frac{1}{2x(1 - x^2) + \sqrt{1 - x^2}}.$$

(17.) $u = \sqrt{a + x + \sqrt{a + x + \sqrt{a + x +}}}$ &c. in inf.

$\qquad u^2 = a + x + \sqrt{a + x + \sqrt{a + x +}}$ &c. in inf.

$\qquad\qquad = a + x + u.$

$$2u\frac{du}{dx}=1+\frac{du}{dx}, \qquad (2u-1)\frac{du}{dx}=1,$$

$$\therefore \frac{du}{dx}=\frac{1}{2u-1}.$$

But $\because u^2-u=a+x,$

$$u^2-u+\overline{\frac{1}{2}\Big|}^2=a+x+\frac{1}{4}=\frac{4a+4x+1}{4},$$

$$u-\frac{1}{2}=\frac{\sqrt{4x+4a+1}}{2}, \quad 2u-1=\sqrt{4x+4a+1},$$

$$\therefore \frac{du}{dx}=\frac{1}{\sqrt{4x+4a+1}}.$$

(18.) $u=1+\dfrac{x}{1+\dfrac{x}{1+\dfrac{x}{1+}}}$ &c. in inf.

$$\therefore u=1+\frac{x}{u}, \qquad u^2-u=x, \qquad 2u\frac{du}{dx}-\frac{du}{dx}=1,$$

$$(2u-1)\frac{du}{ux}=1, \qquad \frac{du}{dx}=\frac{1}{2u-1},$$

$$u^2-u+\overline{\frac{1}{2}\Big|}^2=x+\frac{1}{4}=\frac{4x+1}{4}, \qquad u-\frac{1}{2}=\frac{\sqrt{4x+1}}{2},$$

$$2u-1=\sqrt{4x+1}, \qquad \therefore \frac{du}{dx}=\frac{1}{\sqrt{4x+1}}.$$

(19.) $2ux+au^2-bx^2=0.$ This is an implicit function.

$$2u+2x\cdot\frac{du}{dx}+2au\cdot\frac{du}{dx}-2bx=0, \qquad u+x\frac{du}{dx}+au\frac{du}{dx}-bx=0,$$

$$(au+x)\cdot\frac{du}{dx}=bx-u, \qquad \therefore \frac{du}{dx}=\frac{bx-u}{au+x}.$$

But $\because bx^2-ux=au^2+ux,$ $(bx-u)x=(au+x)u,$

$$\therefore \frac{bx-u}{au+x}=\frac{u}{x};\qquad \therefore \frac{du}{dx}=\frac{u}{x}.$$

(20.) $u=2a+5x.$ $\dfrac{du}{dx}=5.$

(21.) $u=m+nx.$ $\dfrac{du}{dx}=n.$

(22.) $u=c-2x^3.$ $\dfrac{du}{dx}=-6x^2.$

(23.) $u=2x^2-3x+6.$ $\dfrac{du}{dx}=4x-3.$

(24.) $u=4x^3-2x^2+3x.$ $\dfrac{du}{dx}=12x^2-4x+3.$

(25.) $u=(a+bx)x^3.$ $\dfrac{du}{dx}=(4bx+3a)x^2.$

(26.) $u=\sqrt{x^2+a^2}.$ $\dfrac{du}{dx}=\dfrac{x}{\sqrt{x^2+a^2}}.$

(27.) $u=\left(\dfrac{x}{1+x}\right)^n.$ $\dfrac{du}{dx}=\dfrac{nx^{n-1}}{(1+x)^{n+1}}.$

(28.) $u=\dfrac{x^2}{(a+x^3)^2}.$ $\dfrac{du}{dx}=\dfrac{2x(a-2x^3)}{(a+x^3)^3}.$

(29.) $u=\{x^2+(a+x^2)^{\frac{1}{2}}\}^{\frac{1}{2}}.$

$$\frac{du}{dx}=\frac{x}{\sqrt{x^2+\sqrt{a+x^2}}}+\frac{x}{2\sqrt{a+x^2+x^2\sqrt{a+x^2}}}.$$

(30.) $a^2y^2+b^2x^2=a^2b^2.$ $\dfrac{dy}{dx}=\dfrac{b}{a}\cdot\dfrac{x}{\sqrt{a^2-x^2}}.$

(31.) $u=(1+x)\sqrt{1-x}.$ $\dfrac{du}{dx}=\dfrac{1-3x}{2\sqrt{1-x}}.$

(32.) $u=\dfrac{x^3}{\sqrt{(1-x^2)^3}}.$ $\dfrac{du}{dx}=\dfrac{3x^2}{(1-x^2)^{\frac{5}{2}}}.$

OF ONE VARIABLE.

7

(33.) $u = \dfrac{x}{\sqrt{x^2+1}+x}$. $\qquad \dfrac{du}{dx} = \dfrac{2x^2+1}{\sqrt{x^2+1}} - 2x$.

(34.) $u = \sqrt{\dfrac{1-\sqrt{x}}{1+\sqrt{x}}}$. $\quad \dfrac{du}{dx} = -\dfrac{1}{2(1+\sqrt{x})\sqrt{x-x^2}}$.

(35.) $u = (ax^3+b)^2 + (x-b)\sqrt{a^2-x^2}$.

$$\dfrac{du}{dx} = 6ax^2(ax^3+b) + \dfrac{2(a^2+bx-2x^2)}{\sqrt{a^2-x^2}}.$$

(36.) $u = \dfrac{8b}{3a}\cdot x\sqrt{ax-x^2}$. $\qquad \dfrac{du}{dx} = \dfrac{8b}{3a}\cdot\dfrac{3ax-4x^2}{2\sqrt{ax-x^2}}$.

(37.) $u = \dfrac{a^4}{2\sqrt{a^2x^2-x^4}}$. $\qquad \dfrac{du}{dx} = \dfrac{-a^4(a^2-2x^2)}{2x^2(a^2-x^2)^{\frac{3}{2}}}$.

(38.) $u = \dfrac{(x+a)^{\frac{5}{4}}}{(x-a)^{\frac{1}{4}}}$. $\dfrac{du}{dx} = \dfrac{3}{2}\sqrt{\dfrac{x+a}{x-a}} - \dfrac{1}{2}\sqrt{\left(\dfrac{x+a}{x-a}\right)^3}$.

(39.) $u = \dfrac{\sqrt{x^2+1}-x}{\sqrt{x^2+1}+x}$. $\dfrac{du}{dx} = \dfrac{-2}{\sqrt{x^2+1}.(\sqrt{x^2+1}+x)^2}$.

(40.) $u = \dfrac{x\sqrt[3]{a^2+x^2}}{\sqrt{a-x}}$. $\dfrac{du}{dx} = \dfrac{3a^2+4ax-x^2}{6(a^2+x^2)^{\frac{2}{3}}.(a-x)^{\frac{3}{2}}}$.

(41.) $u = x(a^2+x^2)(a^2-x^2)^{\frac{1}{2}}$. $\dfrac{du}{dx} = \dfrac{a^4+a^2x^2-4x^4}{\sqrt{a^2-x^2}}$.

(42.) $u = \sqrt{2x-1-\sqrt{2x-1-\sqrt{2x-1}}}$ —&c. in inf.

$$\dfrac{du}{dx} = \dfrac{2}{\sqrt{8x-3}}.$$

(43.) $u = \dfrac{x^n}{1-\dfrac{x^n}{1-\dfrac{x^n}{1-}}}$ &c. in inf. $\qquad \dfrac{du}{dx} = -\dfrac{nx^{n-1}}{\sqrt{1-4x^n}}$.

(44.) $u = \dfrac{2x^2}{1 + (1 - 4x^n)^{\frac{1}{2}}}$.

$$\frac{du}{dx} = \frac{2x\{1 + (1 - 4x^n)^{\frac{1}{2}} + nx^n(1 - 4x^n)^{-\frac{1}{2}}\}}{1 + (1 - 4x^n)^{\frac{1}{2}} - 2x^n}.$$

CHAPTER II.

TRANSCENDENTAL FUNCTIONS OF ONE VARIABLE.

If $u = \sin x$; $\dfrac{du}{dx} = \cos x.$

$u = \cos x$; $\dfrac{du}{dx} = -\sin x.$

$u = \tan x$; $\dfrac{du}{dx} = 1 + \tan^2 x = \sec^2 x = \dfrac{1}{\cos^2 x}.$

$u = \cot x$; $\dfrac{du}{dx} = -(1 + \cot^2 x) = -\operatorname{cosec}^2 x = -\dfrac{1}{\sin^2 x}.$

$u = \sec x$; $\dfrac{du}{dx} = \sec x . \tan x.$

$u = \operatorname{cosec} x$; $\dfrac{du}{dx} = -\operatorname{cosec} x . \cot x.$

$u = \text{v. } \sin x$; $\dfrac{du}{dx} = \sin x.$

$u = \log ax$; $\dfrac{du}{dx} = \dfrac{a}{x}.$

$u = e^x$; $\dfrac{du}{dx} = e^x.$

Ex. (1.) Let $u = \sin^2 x$. Then $\dfrac{du}{dx} = 2\sin x . \dfrac{d \sin x}{dx} = 2\sin x . \cos x.$

(2.) $\quad u=\cos mx.$ i.e. the cosine of the product of m and x.

$$\frac{du}{dx}=\frac{d\cos mx}{dx}=-\sin mx.\frac{dmx}{dx}=-m\sin mx.$$

(3.) $\quad u=\sin^3 x.\cos x.$

$$\frac{du}{dx}=\sin^3 x.\frac{d\cos x}{dx}+\cos x.\frac{d\sin^3 x}{dx}$$

$$=\sin^3 x.(-\sin x)+\cos x.\,3\sin^2 x.\cos x$$

$$=3\sin^2 x\cos^2 x-\sin^4 x=\sin^2 x\,(3\cos^2 x-\sin^2 x)$$

$$=\sin^2 x\,(3.\overline{1-\sin^2 x}-\sin^2 x)=\sin^2 x(3-3\sin^2 x-\sin^2 x)$$

$$=\sin^2 x\,(3-4\sin^2 x).$$

(4.) $\quad u=e^x.\cos x,$ e being the base of the Napierian system of logarithms.

$$\frac{du}{dx}=e^x.\frac{d\cos x}{dx}+\cos x.\frac{de^x}{dx}$$

$$=e^x.(-\sin x)+\cos x.\,e^x=e^x\,(\cos x-\sin x).$$

(5.) $\quad u=x.\,e^{\cos x}.$

$$\frac{du}{dx}=x.\frac{de^{\cos x}}{dx}+e^{\cos x}.\frac{dx}{dx}=x.\,e^{\cos x}.\frac{d\cos x}{dx}+e^{\cos x}$$

$$=x.\,e^{\cos x}(-\sin x)+e^{\cos x}=e^{\cos x}\,(1-x\sin x).$$

(6.) $\quad u=\dfrac{\sin^m x}{\cos^n x}.$

$$\frac{du}{dx}=\frac{\cos^n x.\,m\sin^{m-1}x.\cos x-\sin^m x.\,n\cos^{n-1}x(-\sin x)}{\cos^{2n}x}$$

$$=\frac{m\cos^{n+1}x.\sin^{m-1}x}{\cos^{2n}x}+\frac{n\sin^{m+1}x.\cos^{n-1}x}{\cos^{2n}x}$$

$$=\frac{m\sin^{m-1}x}{\cos^{n-1}x}+\frac{n\sin^{m+1}x}{\cos^{n+1}x}.$$

(7.) $u = \cos^{-1} x \sqrt{1-x^2}$.   This is an inverse function.

Put $x \sqrt{1-x^2} = z$.   Then $u = \cos^{-1} z$; $\therefore \cos u = z$;

$$-\sin u . \frac{du}{dz} = 1 \quad \therefore \frac{du}{dz} = -\frac{1}{\sin u} = -\frac{1}{\sqrt{1-\cos^2 u}} = -\frac{1}{\sqrt{1-z^2}}$$

$$= -\frac{1}{\sqrt{1-x^2+x^4}}.$$

But $\because z = x \sqrt{1-x^2}$, $\therefore \dfrac{dz}{dx} = x . \dfrac{-x}{\sqrt{1-x^2}} + \sqrt{1-x^2}$

$$= \frac{-x^2 + 1 - x^2}{\sqrt{1-x^2}} = \frac{1-2x^2}{\sqrt{1-x^2}}.$$

Hence $\dfrac{du}{dx} = \dfrac{du}{dz} . \dfrac{dz}{dx} = -\dfrac{1}{\sqrt{1-x^2+x^4}} . \dfrac{1-2x^2}{\sqrt{1-x^2}}$

$$= -\frac{1-2x^2}{\sqrt{(1-x^2+x^4)(1-x^2)}}.$$

(8.) $u = a (\sin x - \cos x)$.

$\dfrac{du}{dx} = a (\cos x + \sin x)$.   Squaring, we have

$$\left(\frac{du}{dx}\right)^2 = a^2(\cos^2 x + \sin^2 x + 2\sin x \cos x) = a^2(1 + 2\sin x \cos x).$$

$$u^2 = a^2(\cos^2 x + \sin^2 x - 2\sin x \cos x) = a^2(1 - 2\sin x \cos x).$$

$$\left(\frac{du}{dx}\right)^2 + u^2 = 2a^2. \quad \therefore \left(\frac{du}{dx}\right)^2 = 2a^2 - u^2. \quad \therefore \frac{du}{dx} = \sqrt{2a^2 - u^2}.$$

(9.) $u = (\log x^n)^m$.

Put z for $\log x^n$, then $u = z^m$, $\therefore \dfrac{du}{dz} = m z^{m-1}$;

and $\because z = \log x^n$, $\therefore \dfrac{dz}{dx} = \dfrac{n x^{n-1}}{x^n} = \dfrac{n}{x}$.

Hence $\dfrac{du}{dx} = \dfrac{du}{dz} . \dfrac{dz}{dx} = m z^{m-1} . \dfrac{n}{x} = \dfrac{m n (\log x^n)^{m-1}}{x}$.

(10.) $\log u = \dfrac{\sqrt{1+x^2}}{x}.$

$$\frac{du}{dx}\cdot\frac{1}{u}=\frac{x\cdot\dfrac{x}{\sqrt{1+x^2}}-\sqrt{1+x^2}}{x^2}=\frac{x^2-\overline{1+x^2}}{x^2\sqrt{1+x^2}}=\frac{-1}{x^2\sqrt{1+x^2}}.$$

$$\therefore \frac{du}{dx}=-\frac{u}{x^2\sqrt{1+x^2}}.$$

(11.) $u=xe^{\tan^{-1}x}.$

$\log u = \log x + \tan^{-1}x . \log e$

$\qquad = \log x + \tan^{-1}x, \quad \because \log e = 1,$

$$\frac{du}{dx}\cdot\frac{1}{u}=\frac{1}{x}+\frac{1}{1+x^2}=\frac{1+x^2+x}{x(1+x^2)}$$

$$=\frac{1+x+x^2}{\dfrac{u}{e^{\tan^{-1}x}}\cdot(1+x^2)}=\frac{e^{\tan^{-1}x}(1+x+x^2)}{u(1+x^2)}.$$

$$\therefore \frac{du}{dx}=\frac{e^{\tan^{-1}x}(1+x+x^2)}{1+x^2}.$$

(12.) $u=\dfrac{e^{ax}(a\sin x-\cos x)}{a^2+1}.$

Since the denominator is constant, and since the differential coefficient of e^{ax} is ae^{ax},

$$\therefore \frac{du}{dx}=\frac{1}{a^2+1}\{ae^{ax}(a\sin x-\cos x)+e^{ax}(a\cos x+\sin x)\}$$

$$=\frac{e^{ax}}{a^2+1}\{a^2\sin x-a\cos x+a\cos x+\sin x\}$$

$$=\frac{e^{ax}}{a^2+1}\cdot(a^2+1)\sin x$$

$$=e^{ax}.\sin x.$$

(13.) $u = \log \dfrac{\sqrt{a} + \sqrt{x}}{\sqrt{a} - \sqrt{x}}.$

$$\frac{du}{dx} = \frac{(\sqrt{a} - \sqrt{x}) \cdot \dfrac{1}{2\sqrt{x}} - (\sqrt{a} + \sqrt{x}) \cdot \left(\dfrac{-1}{2\sqrt{x}}\right)}{(\sqrt{a} - \sqrt{x})^2} \cdot \frac{\sqrt{a} - \sqrt{x}}{\sqrt{a} + \sqrt{x}}$$

$$= \frac{\sqrt{a} - \sqrt{x} + \sqrt{a} + \sqrt{x}}{2\sqrt{x}(\sqrt{a} - \sqrt{x})(\sqrt{a} + \sqrt{x})} = \frac{2\sqrt{a}}{2\sqrt{x}(a - x)} = \frac{\sqrt{a}}{\sqrt{x}(a - x)}.$$

(14.) $u = a^{c^{x^2 + x}}.$

$\log u = c^{x^2 + x} \cdot \log a.$

$\log (\log u) = (x^2 + x) \log c + \log (\log a).$

Put $z = \log u,$ then $\dfrac{dz}{du} = \dfrac{1}{u} = \dfrac{1}{a^{c^{x^2 + x}}},$

$\log z = (x^2 + x) \log c + \log (\log a),$

$\dfrac{dz}{dx} \cdot \dfrac{1}{z} = \log c \,(2x + 1).$ But $z = c^{x^2 + x} \cdot \log a,$

$\therefore \dfrac{dz}{dx} = z . \log c \,(2x + 1) = \log a . \log c . c^{x^2 + x} . (2x + 1).$

Hence $\dfrac{du}{dx} = \dfrac{dz}{dx} . \dfrac{du}{dz} = \log a . \log c . a^{c^{x^2 + x}} . c^{x^2 + x} . (2x + 1).$

(15.) $u = \sin x \sqrt{-1} - \cos x.$ $\dfrac{du}{dx} = \cos x \sqrt{-1} + \sin x.$

(16.) $u = \cos (\sin x).$ $\dfrac{du}{dx} = -\cos x \sin (\sin x).$

(17.) $u = \sin^{-1} \dfrac{x}{a}.$ $\dfrac{du}{dx} = \sqrt{a^2 - x^2}.$

(18.) $u = e^{\sin x}.$ $\dfrac{du}{dx} = e^{\sin x} . \cos x.$

(19.) $u = \sin^{-1} \dfrac{1 - x^2}{1 + x^2}.$ $\qquad \dfrac{du}{dx} = -\dfrac{2}{1 + x^2}.$

(20.) $u = \sin^{-1} \sqrt{\dfrac{x^2 - a^2}{b^2 - a^2}}.$ $\quad \dfrac{du}{dx} = \dfrac{x}{\sqrt{(x^2 - a^2)(b^2 - x^2)}}.$

(21.) $u = \cos x + \cos 2x + \cos 3x + \text{ &c.}$

$$\dfrac{du}{dx} = -\,(\sin x + 2\sin 2x + 3\sin 3x + \text{ &c.}).$$

(22.) $u = \cot^{-1}(mx + a)^2.$ $\qquad \dfrac{du}{dx} = -\dfrac{2m\,(mx + a)}{(mx + a)^4 + 1}.$

(23.) $u = \tan^{-1} \sqrt{\dfrac{1 - x}{1 + x}}.$ $\qquad \dfrac{du}{dx} = -\dfrac{1}{2\sqrt{1 - x^2}}.$

(24.) $u = \tan^{-1} \sqrt{\dfrac{bx + a}{b - a}}.$ $\qquad \dfrac{du}{dx} = \dfrac{\sqrt{b - a}}{2(1 + x)\sqrt{bx + a}}.$

(25.) $u = \sqrt{1 - x^2} + \sin^{-1} x.$ $\qquad \dfrac{du}{dx} = \sqrt{\dfrac{1 - x}{1 + x}}.$

(26.) $u = \operatorname{cosec}^{-1} mx^2.$ $\qquad \dfrac{du}{dx} = \dfrac{-2}{x\sqrt{m^2 x^4 - 1}}.$

(27.) $u = \log(\sin x).$ $\qquad \dfrac{du}{dx} = \cot x.$

(28.) $u = \sin(\log x).$ $\qquad \dfrac{du}{dx} = \dfrac{1}{x}\cos(\log x).$

(29.) $u = \log \dfrac{x}{x + \sqrt{1 + x^2}}.$ $\qquad \dfrac{du}{dx} = \dfrac{1}{x} - \dfrac{1}{\sqrt{1 + x^2}}.$

(30.) $u = (\log)^n x.$ * $\qquad \dfrac{du}{dx} = \dfrac{1}{x \log x\,(\log)^2 x \dots (\log)^{n-1} x}.$

* This expression means the nth logarithm of x, not the nth power of the logarithm of x. Log $(\log x)$, which means the logarithm of the logarithm of x, might be written $(\log)^2 x$.

C

(31.) $\quad u = \log x - \log(a - \sqrt{a^2 - x^2}).\qquad \dfrac{du}{dx} = -\dfrac{a}{x\sqrt{a^2 - x^2}}.$

(32.) $\quad u = \log\sqrt{\sin x} + \log\sqrt{\cos x}.\qquad \dfrac{du}{dx} = \cot 2x.$

(33.) $\quad u = \log\{x + \sqrt{x^2 - a^2}\} + \sec^{-1}\dfrac{x}{a}.\qquad \dfrac{du}{dx} = \dfrac{\sqrt{x + a}}{x\sqrt{x - a}}.$

(34.) $\quad u = \log\dfrac{x\sqrt{2} + \sqrt{1 + x^2}}{\sqrt{1 - x^2}}.\qquad \dfrac{du}{dx} = \dfrac{\sqrt{2}}{(1 - x^2)\sqrt{1 + x^2}}.$

(35.) $\quad u = x^x.\qquad \dfrac{du}{dx} = x^x(\log x + 1).$

(36.) $\quad u = a^{\log x}.\qquad \dfrac{du}{dx} = \dfrac{a^{\log x}.\log a}{x}.$

(37.) $\quad u = x^{x^x}.\qquad \dfrac{du}{dx} = x^{x^x}.x^x\left\{\log x(\log x + 1) + \dfrac{1}{x}\right\}.$

(38.) $\quad u = x^{\sin x}.\qquad \dfrac{du}{dx} = x^{\sin x}\left(\dfrac{\sin x}{x} + \cos x.\log x\right).$

(39.) $\quad u = e^{ax}\cos rx.\qquad \dfrac{du}{dx} = e^{ax}(a\cos rx - r\sin rx).$

(40.) $\quad u = xe^{\tan^{-1}x}.\qquad \dfrac{du}{dx} = e^{\tan^{-1}x}\left(\dfrac{x^2 + x + 1}{x^2 + 1}\right).$

(41.) $\quad u = x^{\frac{1}{x}}.\qquad \dfrac{du}{dx} = \dfrac{1}{x^2}.x^{\frac{1}{x}}\log\left(\dfrac{e}{x}\right).$

(42.) $u = e^{ax}(\sin rx)^m,\quad \dfrac{du}{dx} = e^{ax}(\sin rx)^{m-1}(d\sin rx + mr\cos rx).$

(43.) $\quad u = e^{(\log)^n x}.\qquad \dfrac{du}{dx} = \dfrac{e^{(\log)^n x}}{x\log x(\log)^2 x \ldots (\log)^{n-1}x}.$

(44.) $\quad u = y\tan x^n,\qquad y$ being a function of $x.$

$$\dfrac{du}{dx} = \tan x^n\dfrac{dy}{dx} + nyx^{n-1}\sec^2 x^n.$$

(45.) $u = z^{v^y}$, z, v, and y being functions of x.

$$\frac{du}{dx} = z^{v^y} \cdot v^y \left\{ \log z \cdot \log v \frac{dy}{dx} + \frac{y}{v} \log z \frac{dv}{dx} + \frac{1}{z} \cdot \frac{dz}{dx} \right\}.$$

(46.) $u = xz + \sin z + az \cos z$, $x = a - a \cos z$.

$$\frac{du}{dx} = \left(\frac{2a - x}{x} \right)^{\frac{1}{2}}.$$

CHAPTER III.

SUCCESSIVE DIFFERENTIATION.

Ex. (1).   Let $u = x^n$.

Differentiating, we obtain the *first* differential coefficient,

$$\frac{du}{dx} = nx^{n-1}.$$

Differentiating, we obtain the *second* differential coefficient,

$$\frac{d^2u}{dx^2} = n(n-1)x^{n-2}.$$

Differentiating as before, we have the *third*,

$$\frac{d^3u}{dx^3} = n(n-1)(n-2)x^{n-3}.$$

$$\frac{d^4u}{dx^4} = n(n-1)(n-2)(n-3)x^{n-4}.$$

$$\frac{d^ru}{dx^r} = n(n-1)(n-2)\dots(n-\overline{r-1})^{n-r}.$$

It must be borne in mind that d^2u, d^3u, d^4u, &c. d^ru are merely *symbols;* and that dx^2, dx^3, dx^4, &c. dx^r are *powers* of dx.

(2.) $u = \log x.$

$$\frac{du}{dx} = \frac{1}{x}, \qquad \frac{d^2u}{dx^2} = -\frac{1}{x^2}, \qquad \frac{d^3u}{dx^3} = \frac{2}{x^3},$$

$$\frac{d^4u}{dx^4} = -\frac{2\cdot 3}{x^4}, \qquad \frac{d^5x}{dx^5} = \frac{2\cdot 3\cdot 4}{x^5}, \quad \&c.$$

(3.) $u = a^x.$ $D_x^n u = (\log a)^n a^x$ $\frac{d^5u}{dx^5} = (\log a)^5 a^x.$

(4.) $u = \sin nx.$ $\frac{d^r u}{dx^r} = n^r \sin\left(nx + r\frac{\pi}{2}\right).$

(5.) $u = e^{ax}.$ $D_x^n u = a^n e^{ax}$ $\frac{d^4u}{dx^4} = a^4 e^{ax}.$

(6.) $u = \dfrac{1+x}{1-x}.$ $D_x^n u = 2\dfrac{2\cdot 3 \cdots n}{(1-x)^{n+1}}$ $\dfrac{d^5u}{dx^5} = \dfrac{240}{(1-x)^6}.$

Leibnitz's Theorem, which is useful in finding the differential coefficient of the product of two or more simple functions, may be thus enunciated, u and v being both functions of x,

$$\frac{d^r(uv)}{dx^r} = v\frac{d^r u}{dx^r} + r\frac{dv}{dx}\frac{d^{r-1}u}{dx^{r-1}} + \frac{r(r-1)}{1\cdot 2}\frac{d^2 v}{dx^2}\frac{d^{r-2}u}{dx^{r-2}} + \&c.$$

CHAPTER IV.

TAYLOR'S THEOREM.

This theorem may be thus enunciated.

If $u = f(x)$, and x take the increment h,

$$f(x+h) = u + \frac{du}{dx}h + \frac{d^2u}{dx^2}\frac{h^2}{1\cdot 2} + \frac{d^3u}{dx^3}\frac{h^3}{1\cdot 2\cdot 3} + \cdots \frac{d^n u}{dx^n}\frac{h^n}{1\cdot 2\ldots n}$$
$$+ \&c.$$

This theorem, written according to the notation of Lagrange, is

$$f(x+h)=f(x)+f'(x)h+f''(x)\frac{h^2}{1\cdot2}+f'''(x)\frac{h^3}{1\cdot2\cdot3}+\&c.$$

In using it, if we take n terms of the series, the error we shall commit by leaving out the terms beyond the nth, will lie between the greatest and least values of $f^{(n)}(x+\theta h)\dfrac{h^n}{1\cdot2\cdot3\ldots n}$, which values will depend upon giving to θ various values between (0) its least value, and (1) its greatest.

Maclaurin's Theorem is easily deducible from this.

Ex. (1.) Expand $\cos(x+h)$ in a series of powers of h.

Let $u=\cos x$, then $\dfrac{du}{dx}=-\sin x$, $\dfrac{d^2u}{dx^2}=-\cos x$, $\dfrac{d^3u}{dx^3}=\sin x$, &c.

Whence, substituting these values of u, $\dfrac{du}{dx}$, &c. in Taylor's theorem, we have

$$\cos(x+h)=\cos x-\sin x.\,h-\cos x\cdot\frac{h^2}{1\cdot2}+\sin x\frac{h^3}{1\cdot2\cdot3}+\&c.$$

Cor. By making $x=0$, we have

$$\cos h=1-\frac{h^2}{1\cdot2}+\frac{h^4}{2\cdot3\cdot4}-\&c.$$

(2.) Expand $\sin^{-1}(x+h)$, according to ascending powers of h.

Let $u=\sin^{-1}x$, then $\dfrac{du}{dx}=\dfrac{1}{\sqrt{1-x^2}}=(1-x^2)^{-\frac{1}{2}}$,

$$\frac{d^2u}{dx^2}=-\frac{1}{2}(1-x^2)^{-\frac{3}{2}}(-2x)=x(1-x^2)^{-\frac{3}{2}}=\frac{x}{(1-x^2)^{\frac{3}{2}}}.$$

$$\frac{d^3u}{dx^3}=x\cdot\left\{-\frac{3}{2}(1-x^2)^{-\frac{5}{2}}(-2x)\right\}+(1-x^2)^{-\frac{3}{2}}$$

$$=(1-x^2)^{-\frac{5}{2}}\{3x^2+(1-x^2)\}=\frac{1+2x^2}{(1-x^2)^{\frac{5}{2}}}.$$

Whence, by substitution in the theorem,

$$\sin^{-1}(x+h)=\sin^{-1}x+\frac{h}{(1-x^2)^{\frac{1}{2}}}+\frac{h^2 x}{2(1-x^2)^{\frac{3}{2}}}$$

$$+\frac{h^3(1+2x^2)}{2\cdot3(1-x^2)^{\frac{5}{2}}}+ \&c.$$

(3.)   Expand $\log(x+h)$ by Taylor's theorem.

Let $u=\log x$,   then $\dfrac{du}{dx}=\dfrac{1}{x}$, $\dfrac{d^2u}{dx^2}=-\dfrac{1}{x^2}$, $\dfrac{d^3u}{dx^3}=\dfrac{2}{x^3}$, &c.

Whence, by substitution,

$$\log(x+h)=\log x+\frac{h}{x}-\frac{h^2}{2x^2}+\frac{h^3}{3x^3}-\&c.$$

(4.)   If $u=f(x)$, show that

$$f\left(\frac{x}{1+x}\right)=u-\frac{du}{dx}\cdot\frac{x^2}{1+x}+\frac{d^2u}{dx^2}\cdot\frac{x^4}{2(1+x)^2}$$

$$-\frac{d^3u}{dx^3}\cdot\frac{x^6}{2\cdot3(1+x)^3}+\&c.$$

Let $x+h=\dfrac{x}{1+x}$, then $h=\dfrac{x}{1+x}-x=-\dfrac{x^2}{1+x}$,

$$h^2=\frac{x^4}{(1+x)^2}, \quad h^3=-\frac{x^6}{(1+x)^3}, \quad \&c.$$

$$u=f(x), \quad f(x+h)=f\left(\frac{x}{1+x}\right).$$

Substituting these values in Taylor's theorem, we have

$$f\left(\frac{x}{x+h}\right)=u-\frac{du}{dx}\cdot\frac{x^2}{1+x}+\frac{d^2u}{dx^2}\cdot\frac{x^4}{2(1+x)^2}$$

$$-\frac{d^3u}{dx^3}\cdot\frac{x^6}{2\cdot3(1+x)^3}+\&c.$$

(5.)   If $f(x)=\tan^{-1}x$, and we put $\dfrac{1}{1+x^2}=\sin y$, or $\tan^{-1}x=\dfrac{\pi}{2}-y$, then, $\tan^{-1}(x+h)=\tan^{-1}x+\sin y\,\sin y\,\dfrac{h}{1}$

$$-\sin 2y\,\sin^2 y\,\dfrac{h^2}{2}+\sin 3y\,\sin^3 y\,\dfrac{h^3}{3}-\&c.$$

Now, since h may have any value whatever, put $h=-x$, y being an arc in the first quadrant; then
$$\tan^{-1}(x+h)=\tan^{-1}0=0,$$

$\therefore \tan^{-1}x=\sin y\,\sin y\cdot\dfrac{x}{1}+\sin 2y\,\sin^2 y\cdot\dfrac{x^2}{2}+\sin 3y\,\sin^3 y\dfrac{x^3}{3}+\&c.$

But $\tan^{-1}x=\dfrac{\pi}{2}-y$, and $x=\cot y=\dfrac{\cos y}{\sin y}$,

$\therefore \dfrac{\pi}{2}=y+\sin y\cos y+\dfrac{1}{2}\sin 2y.\cos^2 y+\dfrac{1}{3}\sin 3y\,\cos^3 y+\&c.$

Similarly, putting $h=-\left(x+\dfrac{1}{x}\right)=-\dfrac{1}{\sin y\cos y}$, we have
$$\dfrac{\pi}{2}=\dfrac{\sin y}{\cos y}+\dfrac{1}{2}\dfrac{\sin 2y}{\cos^2 y}+\dfrac{1}{3}\dfrac{\sin 3y}{\cos^3 y}+\&c.$$

And, putting $h=-\sqrt{1+x^2}$,
$$\dfrac{\pi}{2}=\dfrac{y}{2}+\sin y+\dfrac{1}{2}\sin 2y+\dfrac{1}{3}\sin 3y+\&c.$$

Hence, by differentiation,
$$\dfrac{1}{2}+\cos y+\cos 2y+\cos 3y+\&c.=0.$$

These formulæ are deductions of Euler's.

Taylor's theorem may be applied to find approximate roots of equations of the higher degrees.

(6.) Show that Taylor's theorem comprehends the Binomial theorem.

(7.) Expand $\sin(x+h)$ by Taylor's theorem.
$$\sin(x+h)=\sin x+\cos x\cdot\dfrac{h}{1}-\sin x\dfrac{h^2}{2}-\cos x\dfrac{h^3}{2\cdot3}+\&c.$$

(8.) Show, by Taylor's theorem, that

$$(a+x+h)^n=(a+x)^n+n(a+x)^{n-1}h+\frac{n(n-1)}{1\cdot 2}(a+x)^{n-2}h^2$$
$$+\&c.$$

(9.) Show that $\tan(x+h)=\tan x+\sec^2 x\dfrac{h}{1}$

$$+2\sec^2 x\tan x\frac{h}{1\cdot 2}+2\sec^2 x(1+3\tan^2 x)\frac{h^3}{1\cdot 2\cdot 3}+\&c.$$

(10.) If $u=\cot^{-1}x$, show that

$$\cot^{-1}(x+h)=u-\sin u\sin u\frac{h}{1}+\sin^2 u\sin 2u\frac{h^2}{2}-\&c.$$

(11.) If $f(x)=\dfrac{1+x}{1-x}$, prove that

$$\frac{1+x+h}{1-x-h}=\frac{1+x}{1-x}+2\left\{\frac{h}{(1-x)^2}+\frac{h^2}{(1-x)^3}+\frac{h^3}{(1-x)^4}+\&c.\right\}.$$

CHAPTER V.

MACLAURIN'S THEOREM.

This theorem, which is used for the development of a function according to the ascending powers of the variable, may be thus enunciated, U_0, U_1, U_2, U_3, &c. representing the values of u, $\dfrac{du}{dx}$, $\dfrac{d^2u}{dx^2}$, $\dfrac{d^3u}{dx^3}$, &c. when $x=0$,

$$u=U_0+U_1 x+U_2\cdot\frac{x^2}{1\cdot 2}+U_3\cdot\frac{x^3}{2\cdot 3}+U_4\cdot\frac{x^4}{2\cdot 3\cdot 4}+\&c.$$

Cor. $\dfrac{du}{dx}=U_1+2U_2\cdot\dfrac{x}{1\cdot 2}+3U_3\cdot\dfrac{x^2}{2\cdot 3}+4U_4\cdot\dfrac{x^3}{2\cdot 3\cdot 4}+\&c.$

This theorem was first given in Stirling's " Lineæ Tertii

Ordinis Newtonianæ." It is, however, generally attributed to Maclaurin, and is improperly styled " Maclaurin's Theorem."

Ex. (1.) Expand $(a+x)^n$, n being any number whatever, positive or negative, integral or fractional, rational or irrational.

Let $u=(a+x)^n$, whence   if $x=0$, $U_0=a^n$.

$$\frac{du}{dx}=n(a+x)^{n-1}, \quad \cdots \cdots \quad , \quad U_1=na^{n-1}.$$

$$\frac{d^2u}{dx^2}=n(n-1)(a+x)^{n-2} \quad \cdots \quad , \quad U_2=n(n-1)a^{n-2}.$$

$$\frac{d^3u}{dx^3}=n(n-1)(n-2)(a+x)^{n-3} \ , \quad U_3=n(n-1)(n-2)a^{n-3}.$$

        &c.   &c.

Substituting these values of U_0, U_1, &c. for u, $\dfrac{du}{dx}$, &c. in Maclaurin's theorem, we have

$$(a+x)^n=a^n+na^{n-1}x+\frac{n(n-1)}{\cdot 2}a^{n-2}x^2+\frac{n(n-1)(n-2)}{2\cdot 3}a^{n-3}x^3$$

     $+$ &c., which is the *Binomial Theorem.*

(2.) Develop a^x.

Let $u=a^x$, whence   if $x=0$, $U_0=a^0=1$.

$$\frac{du}{dx}=*Aa^x, \quad \cdots \cdots \cdot \quad U_1 \ = \ A.$$

$$\frac{d^2u}{dx^2}=A^2a^x, \quad \cdots \cdots \cdot \quad U_2 \ = \ A^2.$$

$$\frac{d^3u}{dx^3}=A^3a^x, \quad \cdots \cdots \cdot \quad U_3 \ = \ A^3.$$

---

* A is here put for the hyp. log. of base a, that is, for the expression $(a-1)-\dfrac{1}{2}(a-1)^2+\dfrac{1}{3}(a-1)^3-$&c.

Whence, by substitution in Maclaurin's theorem,

$$a^x = 1 + Ax + \frac{A^2 x^2}{1 \cdot 2} + \frac{A^3 x^3}{1 \cdot 2 \cdot 3} + \&c.,$$

which is the *Exponential Theorem*.

$$\because A = \log a, \therefore a^x = 1 + x\log a + \frac{1}{2}(x\log a)^2 + \frac{1}{2 \cdot 3}(x\log a)^3 + \&c.$$

When $x=1$, $a = 1 + \log a + \frac{1}{2}(\log a)^2 + \frac{1}{2 \cdot 3}(\log a)^3 + \&c.$

an expression for any number a, in terms of its Napierian logarithm.

If for a we write the Napierian base e, we have, since $\log e = 1$,

$$e^x = 1 + x + \frac{x^2}{2} + \frac{x^3}{2 \cdot 3} + \&c.$$

And, when $x=1$,

$$e = 1 + 1 + \frac{1}{2} + \frac{1}{2 \cdot 3} + \&c. = 2 \cdot 71828 \ \&c.$$

(3.)   Expand $\tan^{-1}x$ by the method of indeterminate coefficients,

$$u = \tan^{-1}x, \text{ whence if } x=0, \ U_0 = \tan^{-1}0 = 0.$$

$$\frac{du}{dx} = \frac{1}{1+x^2} = 1 - x^2 + x^4 - x^6 + \&c., \text{ by actual division.}$$

But (Maclaurin's Theor. Cor.),

$$\frac{du}{dx} = U_1 + 2U_2 \cdot \frac{x}{2} + 3U_3 \cdot \frac{x^2}{2 \cdot 3} + 4U_4 \cdot \frac{x^3}{2 \cdot 3 \cdot 4} + \&c.$$

$$\therefore U_1 + 2U_2 \cdot \frac{x}{2} + 3U_3 \cdot \frac{x^2}{2 \cdot 3} + 4U_4 \cdot \frac{x^3}{2 \cdot 3 \cdot 4} + 5U_5 \cdot \frac{x^4}{2 \cdot 3 \cdot 4 \cdot 5} + \&c.$$

$$= 1 - x^2 + x^4 - x^6 + \&c. ;$$

Equating coefficients of like powers of x, we have

$$U_1 = 1, \ U_2 = 0, \ U_3 = -2, \ U_4 = 0, \ U_5 = 2 \cdot 3 \cdot 4, \ \&c. ;$$

whence by substitution, $u = x - \dfrac{2x^3}{2\cdot 3} + \dfrac{2\cdot 3\cdot 4x^5}{2\cdot 3\cdot 4\cdot 5} - \&c.$

$\therefore \tan^{-1}x = x - \dfrac{x^3}{3} + \dfrac{x^5}{5} - \dfrac{x^7}{7} + \&c.$

$\because \tan u = x, \therefore u = \tan u - \dfrac{\tan^3 u}{3} + \dfrac{\tan^5 u}{5} - \dfrac{\tan^7 u}{7} + \&c.,$

which is an expression for the arc, in terms of its tangent.

By help of this and Machin's Formula, we may find an approximate expression for the length of the circumference of a circle.

Let $\tan a = \dfrac{1}{5}$, $A = 4a$, then $A = 4\tan^{-1}\dfrac{1}{5}$;

$\tan A = \dfrac{4\tan a - 4\tan^3 a}{1 - 6\tan^2 a + \tan^4 a} = \dfrac{\dfrac{4}{5} - \dfrac{4}{125}}{1 - \dfrac{6}{25} + \dfrac{1}{625}} = \dfrac{120}{119}.$

Now $\tan(A - 45^\circ) = \dfrac{\tan A - 1}{\tan A + 1} = \dfrac{\dfrac{120}{119} - 1}{\dfrac{120}{119} + 1} = \dfrac{1}{239},$

$\therefore A - 45^\circ = \tan^{-1}\dfrac{1}{239};$

$\therefore 45^\circ = A - \tan^{-1}\dfrac{1}{239},$ or $\dfrac{\pi}{4} = 4\tan^{-1}\dfrac{1}{5} - \tan^{-1}\dfrac{1}{239}$*

$= 4\left(\dfrac{1}{5} - \dfrac{1}{3(5)^3} + \dfrac{1}{5(5)^5} - \dfrac{1}{7(7)^7} + \&c.\right)$

$- \left(\dfrac{1}{239} - \dfrac{1}{3(239)^3} + \dfrac{1}{5(239)^5} - \&c.\right),$

* This is Machin's Formula.

a very convergent series, by which, taking seven terms in the first row, and three in the second, we obtain

$$\pi = 3 \cdot 141592653589793,$$

which is the approximate length of the semicircle, the radius being unity. By taking three terms in the first row, and one in the second, we obtain $\pi = 3 \cdot 1416$, an approximation sufficiently near for ordinary purposes.

(4.)   Expand $\sec x$, in ascending powers of x.

Put $u = \sec x$, whence   if $x = 0$, $\sec x = 1$, $U_0 = 1$.

$$\frac{du}{dx} = \sec x \tan x, \quad \ldots \quad \ldots \quad \tan x = 0, \; U_1 = 0.$$

$$\frac{d^2 u}{dx^2} = \sec x (1 + \tan^2 x) + \tan x \sec x \tan x$$

$$= \sec x + 2 \sec x \tan^2 x, \quad \ldots \quad \ldots \quad U_2 = 1.$$

$$\frac{d^3 u}{dx^3} = \sec x \tan x + 2 \sec x . 2 \tan x (1 + \tan^2 x) + 2 \tan^2 x \sec x \tan x$$

$$= 5 \sec x \tan x + 6 \sec x \tan^3 x, \quad \ldots \quad . \quad U_3 = 0.$$

$$\frac{d^4 x}{dx^4} = 5 \sec x (1 + \tan^2 x) + 5 \tan x \sec x \tan x$$

$$+ 6 \sec x . 3 \tan^2 x (1 + \tan^2 x) + 6 \tan^3 x \sec x \tan x$$

$$= 5 \sec x + 28 \sec x \tan^2 x + 24 \sec x \tan^4 x, . \; . \; U_4 = 5.$$

Whence, by substitution,

$$u = \sec x = 1 + \frac{x^2}{2} + \frac{5 x^4}{2 \cdot 3 \cdot 4} + \&c.$$

(5.)   Expand $\cos^3 x$.

Put $u = \cos^3 x$, whence if $x = 0$, $\cos^3 x = 1$,   . $U_0 = 1$.

$$\frac{du}{dx} = 3 \cos^2 x (- \sin x) = 3 \sin^3 x - 3 \sin x, \; . \; . \; U_1 = 0.$$

$$\frac{d^2 u}{dx^2} = 9 \sin^2 x \cos x - 3 \cos x, \; . \; . \; . \; . \; . \; U_2 = -3.$$

$$\frac{d^3u}{dx^3}=9\sin^2x\,(-\sin x)+\cos x.\,18\sin x\,\cos x+3\sin x$$

$$=3\sin x-9\sin^3x+18\sin x\cos^2x,\quad \ldots \quad U_3=0.$$

$$\frac{d^4u}{dx^4}=3\cos x-27\sin^2x\cos x+18\sin x.\,2\cos x\,(-\sin x)$$

$$+18\cos^2x.\cos x,\quad \ldots\ \ldots\ \ldots\ \ldots\quad U_4=21.$$

$$\therefore u=\cos^3x=1-\frac{3x^2}{2}+\frac{21x^4}{2\cdot3\cdot4}-\&c.=1-\frac{3x^2}{2}+\frac{7x^4}{8}-\&c.$$

(6.) Develop $(1+e^x)^n$ according to ascending powers of x. Let $u=(1+e^x)^n$, whence if $x=0$, $(1+e^0)^n=(1+1)^n$,

$$U_0=2^n.$$

$$\frac{du}{dx}=n\,(1+e^x)^{n-1}.e^x,\quad \ldots\ \ldots\ \ldots\ \ldots\quad U_1=n\,2^{n-1}.$$

$$\frac{d^2u}{dx^2}=n\,(1+e^x)^{n-1}e^x+e^x.n(n-1)(1+e^x)^{n-2}e^x;\ \text{make } x=0,$$

$$\frac{d^2u}{dx^2}=n\,2^{n-1}\qquad +\quad n\,(n-1)2^{n-2},\ \ U_2=n\,2^{n-2}(n+1).$$

$$\frac{d^3u}{dx^3}=n\,(1+e^x)^{n-1}e^x+e^x.n\,(n-1)(1+e^x)^{n-2}.e^x$$

$$+n(n-1)(1+e^x)^{n-2}.e^{2x}.2+e^{2x}.n(n-1)(n-2)(1+e^x)^{n-3}.e^x\ ;$$
$$\text{make } x=0,$$

$$\frac{d^3u}{dx^3}=n\,2^{n-2}(n+1)+n\,(n-1)\,2^{n-2}.2+n(n-1)(n-2).2^{n-3}$$

$$=n\,(n+1)\,2^{n-2}+n\,(n-1)\,2^{n-1}+n\,(n-1)\,(n-2)\,2^{n-3}$$

$$=n\,2^{n-3}\{(n+1)2+(n-1)2^2+(n-1)(n-2)\}$$

$$=n\,2^{n-3}\{2n+2+4n-4+n^2-3n+2\}$$

$$=n\,2^{n-3}\{n^2+3n\},\quad \ldots\ \ldots\ \ldots\quad U_3=n^2\,2^{n-3}(n+3).$$

Whence, by substitution in the theorem,

$$(1+e^x)^n=2^n\Big\{1+\frac{n}{2}\frac{x}{1}+\frac{n(n+1)}{2^2}\,\frac{x^2}{1\cdot2}+\frac{n^2(n+3)}{2^3}\,\frac{x^3}{1\cdot2\cdot3}+\&c.\Big\}$$

D

(7.) Prove that $\log(1+x)=x-\dfrac{x^2}{2}+\dfrac{x^3}{3}-\dfrac{x^4}{4}+\&c.$

Let $u=\log(1+x)$, whence if $x=0$, $U_0=\log(1)=0.$

$\dfrac{du}{dx}=\dfrac{1}{1+x}=1-x+x^2-x^3+x^4-\&c.$ by actual division.

But (Maclaurin's Theorem. Cor.)

$$\dfrac{du}{dx}=U_1+U_2x+\dfrac{U_3}{2}x^2+\dfrac{U_4}{2\cdot3}x^3+\dfrac{U_5}{2\cdot3\cdot4}x^4+\&c.$$

And, equating coefficients of like powers of x,

$$U_1=1,\ \ U_2=-1,\ \ \dfrac{U_3}{2}=1,\ \ \dfrac{U_4}{2\cdot3}=-1,\ \ \dfrac{U_5}{2\cdot3\cdot4}=1\ ;$$

$$\therefore U_1=1,\ \ U_2=-1,\ \ U_3=2,\ \ U_4=-2\cdot3,\ \ U_5=2\cdot3\cdot4.$$

Whence, by substitution in the theorem,

$$\log(1+x)=x-\dfrac{x^2}{2}+\dfrac{x^3}{3}-\dfrac{x^4}{4}+\&c.$$

Cor. Writing $-x$ for x we have

$$\log(1-x)=-x-\dfrac{x^2}{2}-\dfrac{x^3}{3}-\dfrac{x^4}{4}-\&c.$$

(8.) Show, by help of the last example, that

$$\log\left(\dfrac{x}{x-1}\right)=\dfrac{1}{x-1}-\dfrac{1}{2}\dfrac{1}{(x-1)^2}+\dfrac{1}{3}\dfrac{1}{(x-1)^3}-\&c.$$

Put $\dfrac{x}{x-1}=1-z,$ then

$$\log(1+z)=z-\dfrac{1}{2}z^2+\dfrac{1}{3}z^3-\&c.\qquad\text{(Ex. 7.)}$$

But $z=\dfrac{x}{x-1}-1=\dfrac{x-x+1}{x-1}=\dfrac{1}{x-1},$

$$\therefore\log\left(\dfrac{x}{x-1}\right)=\dfrac{1}{x-1}-\dfrac{1}{2}\dfrac{1}{(x-1)^2}+\dfrac{1}{3}\dfrac{1}{(x-1)^3}-\&c.$$

(9.) If a_n and b_n respectively represent the coefficients of x^n in the expansions of $u = f(x)$, and $\log u$; show that
$$na_n = b_1 a_{n-1} + 2b_2 a_{n-2} + 3b_3 a_{n-3} + \dots + nb_n a_0.$$

Assume $u = a_0 + a_1 x + a_2 x^2 \dots + a_n x^n$, then
$$\frac{du}{dx} = a_1 + 2a_2 x \dots + na_n x^{n-1},$$

$$\frac{du}{dx} \cdot \frac{1}{u} = \frac{a_1 + 2a_2 x + 3a_3 x^2 \dots + na_n x^{n-1}}{a_0 + a_1 x + a_2 x^2 \dots + a_n x^n} = \text{diff. coeff. of } \log u.$$

Now $\log u = b_0 + b_1 x + b_2 x^2 \dots + b_n x^n$
$$\therefore \frac{du}{dx} \cdot \frac{1}{u} = b_1 + 2b_2 x \dots + nb_n x^{n-1}$$

Hence $\dfrac{a_1 + 2a_2 x \dots + na_n x^{n-1}}{a_0 + a_1 x + a_2 x^2 \dots + a_n x^n} = b_1 + 2b_2 x \dots + nb_n x^{n-1}.$

And, multiplying by the denominator, and equating coefficients of like powers of x, we have
$$na_n = b_1 a_{n-1} + 2b_2 a_{n-2} + 3b_3 a_{n-3} + \dots + nb_n a_0.$$

(10.) Develop $\sin x$ and $\cos x$ in ascending powers of x.
$$\sin x = x - \frac{x^3}{1 \cdot 2 \cdot 3} + \frac{x^5}{1 \cdot 2 \cdot 3 \cdot 4} - \&c.$$
$$\cos x = 1 - \frac{x^2}{1 \cdot 2} + \frac{x^4}{1 \cdot 2 \cdot 3 \cdot 4} - \&c.$$

(11.) Prove Euler's formulæ,
$$\sin x = \frac{e^{x\sqrt{-1}} - e^{-x\sqrt{-1}}}{2\sqrt{-1}}, \quad \cos x = \frac{e^{x\sqrt{-1}} + e^{-x\sqrt{-1}}}{2}.$$

(12.) Prove De Moivre's formula,
$$\cos mx + \sqrt{-1} \sin mx = (\cos x + \sqrt{-1} \sin x)^m.$$

(13.) Prove that $(\tan x)^4 = x^4 + \frac{4}{3} x^6 + \frac{6}{5} x^8 + \&c.$

(14.) If $u = \sin^{-1} x$, show that
$$u = \sin u + \frac{\sin^3 u}{2 \cdot 3} + \frac{3^2 \sin^5 u}{2 \cdot 3 \cdot 4 \cdot 5} + \frac{3^2 \cdot 5^2 \sin^7 u}{2 \cdot 3 \cdot 4 \cdot 5 \cdot 6 \cdot 7} + \&c.$$

(15.) Develop $u = \cot x$ by the method of indeterminate coefficients.

$$\cot x = \frac{1}{x} - \frac{x}{3} - \frac{x^3}{3^2 \cdot 5} - \frac{2x^5}{3^3 \cdot 5 \cdot 7} - \&c.$$

(16.) Prove, by Maclaurin's theorem, that

$$(1 + 2x + 3x^2)^{-\frac{1}{2}} = 1 - x + 2x^3 - \frac{7}{2}x^4 + \frac{3}{2}x^5 - \&c.$$

(17.) Show that $\cos^{-1} x = \dfrac{\pi}{2} - x - \dfrac{x^3}{2 \cdot 3} - \dfrac{3^2 x^5}{2 \cdot 3 \cdot 4 \cdot 5} - \&c.$

(18.) Show that $\sin(a + bx + cx^2) = \sin a + bx \cos a$

$$+ \frac{2c \cos a - b^2 \sin a}{2} x^2 - \frac{6bc \sin a + b^3 \cos a}{2 \cdot 3} x^3 - \&c.$$

(19.) Prove that $\dfrac{x^2}{e^x - x} = \dfrac{x^2}{1} - \dfrac{x^4}{1 \cdot 2} - \dfrac{x^5}{1 \cdot 2 \cdot 3} - \&c.$

(20.) If $\cos x + \sin x \sqrt{-1} = e^{x\sqrt{-1}}$, and x take the particular value $\dfrac{\pi}{2}$, prove the two formulæ of John Bernouilli, namely,

$$\pi = -\sqrt{-1} . \log(-1), \quad \text{and}$$

$$(\sqrt{-1})^{\sqrt{-1}} = 1 - \frac{\pi}{2} + \frac{1}{1 \cdot 2}\left(\frac{\pi}{2}\right)^2 - \frac{1}{1 \cdot 2 \cdot 3}\left(\frac{\pi}{2}\right)^3 + \&c.$$

Implicit Functions.

Ex. (1.) Given $u^3 - 3u + x = 0$, to expand u in a series of ascending powers of x.

When $x = 0$, $u^3 - 3u = 0$, $\therefore u = 0$, ... $\therefore U_0 = 0$.

$$3u^2 \frac{du}{dx} - 3\frac{du}{dx} + 1 = 0, \quad \frac{du}{dx} = -\frac{1}{3} \cdot \frac{1}{u^2 - 1}, \quad \ldots \quad U_1 = \frac{1}{3}.$$

$$\frac{d^2u}{dx^2} = -\frac{1}{3}\frac{-2\,u\frac{du}{dx}}{(u^2-1)^2} = \frac{2}{3}\cdot\frac{u}{(u^2-1)^2}\cdot\frac{-1}{3(u^2-1)}$$

$$= -\frac{2}{9}\frac{u}{(u^2-1)^3}, \quad \ldots \ldots \ldots \ldots \quad U_2 = 0.$$

$$\frac{d^3u}{dx^3} = -\frac{2}{9}\frac{(u^2-1)^3\frac{du}{dx} - u.\,3(u^2-1)^2.\,2\,u\frac{du}{dx}}{(u^2-1)^6}$$

$$= -\frac{2}{9}\cdot\frac{-5u^2-1}{(u^2-1)^4}\cdot\frac{-1}{3(u^2-1)} = -\frac{2}{27}\frac{5u^2+1}{(u^2-1)^5}, \ldots\; U_3 = \frac{2}{27}.$$

$$\frac{d^4u}{dx^4} = -\frac{2}{27}\cdot\frac{(u^2-1)^5\,10\,u\frac{du}{dx} - (5u^2+1).5(u^2-1)^4.\,2\,u\frac{du}{dx}}{(u^2-1)^{10}}$$

$$= \frac{20}{81}\cdot\frac{-4u^3-2u}{(u^2-1)^7} = -\frac{48}{81}\cdot\frac{2u^3+u}{(u^2-1)^7}, \;\ldots\ldots\; U_4 = 0.$$

$$\frac{d^5u}{dx^5} = -\frac{40}{243}\cdot\frac{22u^4+19u^2+1}{(u^2-1)^9}, \;\ldots\ldots\ldots\; U_5 = \frac{40}{243}.$$

Whence, by substitution in Maclaurin's theorem,

$$u = \frac{x}{3} + \frac{x^3}{3^4} + \frac{x^5}{3^6} + \&\text{c}.$$

(2.) $2u^3 - ux - 2 = 0$; expand u in a series of ascending powers of x.

$$u = 1 + \frac{x}{2\cdot3} - \frac{x^3}{2^3\cdot3^4} + \&\text{c}.$$

(3.) $u^2 - \dfrac{8}{u} = 6x$; show that $u = 2 + x - \dfrac{1}{2}\dfrac{x^3}{2\cdot3} + \dfrac{x^4}{2\cdot3\cdot4} + \&\text{c}.$

(4.) $u^3x - 8u - 8x = 0$; show that $u = -x - \dfrac{x^4}{2^3} - \dfrac{3\,x^7}{2^6} - \&\text{c}.$

(5.) $4u^3x - u - 4 = 0$; show that $u = -4 - 4^4x - 3(4)^7x^2 - \&\text{c}.$

(6.) $u^3 - a^2u + aux - x^3 = 0$; show that

$$u = -\frac{x^3}{a^2} - \frac{x^4}{a^3} - \frac{x^5}{a^4} - \&c.$$

(7.) $\sin u = x \sin(a + u)$,   show that

$$u = r\pi + \sin a \frac{x}{1} + \sin 2a \frac{x^2}{1 \cdot 2} + 2\sin a\,(3 - 4\sin^2 u)\,\frac{x^3}{1 \cdot 2 \cdot 3} + \&c.$$

CHAPTER VI.

EVALUATION OF INDETERMINATE FUNCTIONS.

When the two terms of any fraction $\dfrac{P}{Q}$ contain a common factor, as $x - a$, and the particular value a be given to x, then, since $x - a$ will be equal to 0, the fraction will assume the form $\dfrac{0}{0}$, and be indeterminate.

Such a fraction is improperly termed a *vanishing fraction;* since its values may be finite, infinite, or nothing.

When the common factor is obvious by inspection, it may of course be removed by division.

The method of John Bernouilli is to differentiate the numerator and denominator, *separately,* until they do not vanish simultaneously by making $x = a$, and thus to determine the true value of the fraction in that case.

If the fraction be of the form $\dfrac{P(x-a)^m}{Q(x-a)^n}$, and m or n be a fraction, this method of successive differentiation will not apply, since, however often we differentiate, we shall never eliminate the common factor.

In this case we may put $a \pm h$ for x, expand both terms of the fraction in a series of ascending powers of h, and then put $h = 0$.

The process of evaluation of indeterminate functions enables us to find the sum of a series for a particular value of the variable.

Ex. (1.) Find the real value of the fraction

$$\frac{ax^2 - 2acx + ac^2}{bx^2 - 2bcx + bc^2} \quad \text{when } x = c.$$

Here $P = ax^2 - 2acx + ac^2$, $\quad Q = bx^2 - 2bcx + bc^2$,

$$\therefore \frac{dP}{dx} = 2ax - 2ac = 0 \quad \text{if } x = c$$

$$\frac{dQ}{dx} = 2bx - 2bc = 0 \quad \text{if } x = c$$

$$\left.\begin{array}{l} \dfrac{d^2P}{dx^2} = 2a \\[2mm] \dfrac{d^2Q}{dx^2} = 2b \end{array}\right\} \quad \therefore \text{ the fraction } = \frac{2a}{2b} = \frac{a}{b}.$$

(2.) Let $u = \dfrac{x^3 + 2x^2 - x - 2}{x^3 - 1}$. Find u, when $x = 1$.

Here $P = x^3 + 2x^2 - x - 2$, $\quad Q = x^3 - 1$;

$$\left.\begin{array}{ll} \dfrac{dP}{dx} = 3x^2 + 4x - 1 = 6 & \text{if } x = 1 \\[2mm] \dfrac{dQ}{dx} = 3x^2 \qquad\quad = 3 & \text{if } x = 1 \end{array}\right\} \quad \therefore u = \frac{6}{3} = 2.$$

(3.) $u = \dfrac{e^x - e^{\sin x}}{x - \sin x} = 1$, $\quad$ when $x = 0$.

$$\frac{dP}{dx} = e^x - e^{\sin x} \cdot \cos x = 1 - 1 = 0 \quad \text{if } x = 0,$$

$$\frac{dQ}{dx} = 1 - \cos x = 1 - 1 = 0 \quad \text{if } x = 0,$$

$$\frac{d^2P}{dx^2}=e^x-e^{\sin x}(-\sin x)-\cos x\, e^{\sin x}.\cos x=0 \text{ if } x=0,$$

$$\frac{d^2Q}{dx^2}=\sin x=0 \quad \text{ if } x=0,$$

$$\frac{d^3P}{dx^3}=e^x+e^{\sin x}\cos x+\sin x\, e^{\sin x}.\cos x-\cos^2 x\, e^{\sin x}.\cos x$$

$$+e^{\sin x}.2\cos x\sin x=1+1+0-1+0=1 \quad \text{ if } x=0,$$

$$\frac{d^3Q}{dx^3}=\cos x=1 \quad \text{ if } x=0, \qquad \therefore u=\frac{1}{1}=1.$$

(4.) $u=(1-x)\tan\dfrac{\pi x}{2}=\dfrac{1-x}{\cot\dfrac{\pi x}{2}}=\dfrac{2}{\pi},$ \qquad when $x=1$.

Here $P=1-x, \qquad Q=\cot\dfrac{\pi}{2}x,$

$$\frac{dP}{dx}=-1. \qquad \frac{dQ}{dx}=-\frac{\dfrac{\pi}{2}}{\sin^2\dfrac{\pi}{2}x}, \qquad \text{make } x=1, \text{ then}$$

$$\frac{dQ}{dx}=-\frac{\dfrac{\pi}{2}}{\sin^2\dfrac{\pi}{2}}=-\frac{\dfrac{\pi}{2}}{1}=-\frac{\pi}{2}. \qquad \therefore u=\frac{-1}{-\dfrac{\pi}{2}}=\frac{2}{\pi}.$$

(5.) $u=\dfrac{(a^2-x^2)^{\frac{1}{2}}+a-x}{(a^3-x^3)^{\frac{1}{3}}+(a-x)^{\frac{1}{3}}}.$ \qquad Find u, when $x=a$.

Put $x=a-h,$ \quad then

$$u=\frac{\{a^2-(a-h)^2\}^{\frac{1}{2}}+a-(a-h)}{\{a^3-(a-h)^3\}^{\frac{1}{3}}+\{a-(a-h)\}^{\frac{1}{3}}}$$

$$=\frac{\{2ah-h^2\}^{\frac{1}{2}}+h}{\{3a^2h-3ah^2+h^3\}^{\frac{1}{3}}+h^{\frac{1}{3}}},$$

$$\therefore u = \frac{h^{\frac{1}{2}}(2a-h)^{\frac{1}{2}}+h}{h^{\frac{1}{2}}(3a^2-3ah+h^2)^{\frac{1}{2}}+h^{\frac{1}{2}}} = \frac{(2a-h)^{\frac{1}{2}}+h^{\frac{1}{2}}}{(3a^2-3ah+h^2)^{\frac{1}{2}}+1}.$$

Now, putting $h=0$, we have $u = \dfrac{(2a)^{\frac{1}{2}}}{3^{\frac{1}{2}}a+1}$.

(6.) $u = \dfrac{\tan x - \sin x}{x^3} = \dfrac{1}{2}$, when $x=0$.

$$\frac{\dfrac{dP}{dx}}{\dfrac{dQ}{dx}} = \frac{\sec^2 x - \cos x}{3x^2} = \frac{1}{\cos^2 x} \cdot \frac{1-\cos^3 x}{3x^2} = \frac{1-\cos^3 x}{3x^2},$$

since the factor $\dfrac{1}{\cos^2 x}=1$ when $x=0$;

$$\frac{d^2P}{dQ^2} = \frac{3\cos^2 x.\sin x}{6x} = \frac{\sin x}{2x}, \quad \because \cos^2 x = 1, \quad \text{when } x=0;$$

$$\frac{d^3P}{dQ^3} = \frac{\cos x}{2} = \frac{1}{2}, \quad \text{hence} \quad \frac{\tan x - \sin x}{x^3} = \frac{1}{2}, \quad \text{when } x=0.$$

(7.) Find the real value of $\dfrac{x^3-3x+2}{x^4-6x^2+8x-3}$ when $x=1$.

Ans. ∞.

(8.) If $u = \dfrac{a-(a^2-x^2)^{\frac{1}{2}}}{x^2}$, when $x=0$, $u = \dfrac{1}{2a}$.

(9.) $u = \dfrac{x^2-a^{\frac{3}{2}}x^{\frac{1}{2}}}{\sqrt{ax}-a}$, when $x=a$, $u=3a$.

(10.) $u = \dfrac{x^3-a^2x-ax^2+a^3}{x^2-a^2}$, when $x=a$, $u=0$.

(11.) $u = \dfrac{x^2-2x}{x^4-4x^3+8x-16}$, when $x=2$, $u=\infty$.

(12.) $u = \dfrac{\cot x + \operatorname{cosec} x - 1}{\cot x - \operatorname{cosec} x + 1}$, when $x = \dfrac{\pi}{2}$, $u=1$.

(13.) $u = \dfrac{x \sin x - \dfrac{\pi}{2}}{\cos x}$, when $x = \dfrac{\pi}{2}$, $u = -1$.

(14.) $u = \dfrac{(x-a)^{\frac{1}{2}} + x^{\frac{1}{2}} - a^{\frac{1}{2}}}{(x^2 - a^2)^{\frac{1}{2}}}$,   when $x = a$, $u = \dfrac{1}{\sqrt{2a}}$.

(15.) If in $\dfrac{1 - x^n}{1 - x}$ $x = 1$, show that $1 + x + x^2 + \ldots x^{n-1} = n$.

(16.) $u = \dfrac{a^m}{1 + x^p} \cdot \dfrac{1 - x^n}{1 - x^p}$,   when $x = 1$, $u = \dfrac{n}{2p}$.

(17.) $u = \dfrac{\sec x (e^{2x} + e^{-2x} - 2)}{x^2}$,   when $x = 0$, $u = 4$.

(18.) $u = \dfrac{e^x - e^{-x}}{\log(1 + x)}$,   when $x = 0$, $u = 2$.

(19.) $u = \dfrac{\tan \pi x - \pi x}{2 x^2 \tan \pi x}$,   when $x = 0$, $u = \dfrac{\pi^2}{6}$.

(20.) $u = \dfrac{a^{\log x} - x}{\log x}$,   when $x = 1$, $u = \log\left(\dfrac{a}{e}\right)$.

(21.) $u = \dfrac{\cos^{-1}(1 - x)}{\sqrt{2x - x^2}}$,   when $x = 0$, $u = 1$.

(22.) $u = \dfrac{x - a + \sqrt{2ax - 2a^2}}{\sqrt{x^2 - a^2}} = 1$,   when $x = a$.

(23.) $u = \dfrac{x^n}{e^x}$,   when $x = \infty$, $u = 0$.

(24.) $u = \dfrac{\log x \cdot (1 - x)}{x - 1 - x \log x}$,   when $x = 1$, $u = 2$.

(25.) $u = x e^{-x} = 0$,   when $x = \infty$.

(26.) $u = \dfrac{\dfrac{1}{4}\dfrac{\pi}{x}}{\cot \dfrac{\pi x}{2}}$,   when $x = 0$, $u = \dfrac{\pi^2}{8}$.

(27.) If $y e^x - x = 0$; show that when x approaches ∞ the limiting values of e^{-x} and y are identical, and that the limiting value of y is zero.

(28.) $u = \dfrac{\log (\tan 2x)}{\log (\tan x)}$,   when $x = 0$, $u = 1$.

(29.) $u = \dfrac{(x^2 + a^2)(x - a)}{x^n - a^n} = \dfrac{2}{n\,a^{n-3}}$,   when $x = a$.

(30.) $u = \dfrac{x^2 - a^2}{x^2} \tan \dfrac{\pi x}{2a}$,   when $x = a$, $u = -\dfrac{4}{\pi}$.

(31.) $u = \dfrac{e^{mx} - e^{ma}}{x - a}$,   when $x = a$, $u = m\,e^{ma}$.

(32.) $u = \dfrac{(x + 1)(x - 1)^{\frac{3}{2}}}{(x - 1)^2 + \sin^3(x^2 - 1)^{\frac{1}{2}}} = \sin 45°$,   when $x = 1$.

(33.) $u = e^{\frac{\log (1 + nx)}{x}}$,   when $x = 0$, $u = e^n$.

(34.) $u = \dfrac{\sin^2 x \cos x}{1 - \cos x}$,   when $x = 0$, $u = 2$.

(35.) If the fraction $\dfrac{1}{f(x)} - \dfrac{1}{\phi(x)}$ assume the form $\infty - \infty$ when $x = a$; show that this illusory form $\infty - \infty$, and also $0 \times \infty$ are each identical with the form $\dfrac{0}{0}$.

CHAPTER VII.

MAXIMA AND MINIMA.

ONE VARIABLE.

If a quantity increases to a certain extent, and then decreases to a certain extent, its values at these limits respectively are a maximum and a minimum.

If it repeatedly increases and decreases alternately, it has several maxima and minima.

If it increases continually or decreases continually, it has no maxima or minima.

Let $u = f(x)$; then, to determine the values of x which render u a maximum or minimum, put $\frac{du}{dx} = 0$ or ∞, and substitute the possible roots of the resulting equation in $\frac{d^2u}{dx^2}$, then, if $\frac{d^2u}{dx^2} = $ a negative quantity, the value of x which is substituted renders u a maximum; if $\frac{d^2u}{dx^2} = $ a positive quantity, the value of x which is substituted renders u a minimum.

A maximum or minimum can exist only when the first differential coefficient which does not vanish is of an *even* order.

If $u = $ a maximum or minimum, then au and $\frac{u}{a}$ are maxima or minima. Hence, before differentiating, we may reject any constant positive factor in the value of u.

If $u = $ a maximum or minimum, then u^n is a maximum or minimum if n is positive; but when $u = $ a maximum u^{-n} is a minimum, and when $u = $ a minimum u^{-n} is a maximum. Hence, before differentiating, we may reject a constant exponent. .

If $u = $ a maximum or minimum, $\log u$ is a maximum or minimum. Hence, when the function consists of a product or quotient of powers or roots, we may use the logarithms.

Ex. (1.) Find when $x^5 - 5x^4 + 5x^3 + 1$ is either a maximum or a minimum.

Let $u = x^5 - 5x^4 + 5x^3 + 1$, then

$$\frac{du}{dx} = 5x^4 - 20x^3 + 15x^2, \quad \text{and putting this} = 0,$$

$$x^2(x^2-4x+3)=0, \quad x=0, \quad x^2-4x+3=0,$$

$$x^2-4x=-3, \quad x=3, \quad x=1,$$

$\dfrac{d^2u}{dx^2}=20x^3-60x^2+30x,$ and substituting successively

the values of x, (0, 1, 3) in this expression,

$\dfrac{d^2u}{dx^2}=0,$ from which we can infer nothing,

$\dfrac{d^2u}{dx^2}=20-60+30=-10,$ which indicates a maximum,

$\dfrac{d^2u}{dx^2}=540-540+90=+90,$ which indicates a minimum.

Hence, when $x=1$, $u=2$, a maximum,

and, when $x=3$, $u=-26$, a minimum.

(2.) If $u=\sqrt{4a^2x^2-2ax^3}$, ascertain those values of x which make u a maximum or minimum.

Rejecting the radical and the common factor $2a$, put

$$u=2ax^2-x^3, \quad \dfrac{du}{dx}=4ax-3x^2=(4a-3x)\,x=0,$$

$$\therefore 4a-3x=0, \quad x=0, \quad \therefore x=\dfrac{4a}{3}, \quad x=0,$$

$$\dfrac{d^2u}{dx^2}=4a-6x=4a-8a=-4a,$$

$$\dfrac{d^2u}{dx^2}=4a-6x=+4a.$$

Hence $x=\dfrac{4a}{3}$ makes $u=\sqrt{\dfrac{64a^4}{9}-\dfrac{128a^4}{27}}=\sqrt{\dfrac{64a^4}{9\times3}}$

$$=\dfrac{8a^2}{3\sqrt{3}}, \text{ a maximum,}$$

$x=0$ makes $u=0$, a minimum.

E

(3.) Determine the maxima and minima values of the function $u = \dfrac{x^2}{\sqrt{1+x^2}}$.

Putting $u = \dfrac{1}{v}$, we shall have $v = \dfrac{1+x^2}{x}$,

$$\frac{dv}{dx} = \frac{x\,2x - (1+x^2)}{x^2} = \frac{x^2-1}{x^2} = 0, \quad \therefore x = +1,\ x = -1,$$

$$\frac{d^2v}{dx^2} = \frac{x^2.2x - (x^2-1)\,2x}{x^4} = \frac{2x}{x^4} = \frac{2}{x^3}.$$

$$\therefore \frac{d^2v}{dx^2} = +\frac{2}{1},\ \text{which indicates a minimum,}$$

$$\frac{d^2v}{dx^2} = -\frac{2}{1}. \quad . \quad . \quad . \quad . \quad . \quad \text{maximum,}$$

$$\therefore u = \frac{1}{1+1} = \frac{1}{2},\ \text{a maximum,}$$

$$u = \frac{-1}{1+1} = -\frac{1}{2},\ \text{a minimum.}$$

(4.) Divide a number a into two such parts that the product of the mth power of the one and the nth power of the other shall be the greatest possible.

Let x, and $a-x$ be the parts, then

$$u = x^m (a-x)^n.$$

$$\frac{du}{dx} = x^m n (a-x)^{n-1}(-1) + (a-x)^n m x^{m-1}$$

$$= x^{m-1} (a-x)^{n-1} \{ -xn + (a-x)\, m \}$$

$$= x^{m-1} (a-x)^{n-1} \{ ma - (m+n)\, x \} = 0 ;$$

$$\therefore x = 0, \qquad x = a, \qquad x = \frac{ma}{m+n}.$$

Or thus, $\log u = m \log x + n \log (a-x)$,

$$\frac{du}{dx}\frac{1}{u} = \frac{m}{x} - \frac{n}{a-x}, \qquad \frac{du}{dx} = u \left(\frac{am - mx - nx}{(a-x)\,x} \right),$$

$$\therefore \frac{du}{dx} = x^m (a-x)^n \cdot \frac{am-(m+n)x}{(a-x)x} = 0,$$

$$\therefore x=0, \qquad x=a, \qquad x=\frac{am}{m+n}.$$

Now the values 0 and a may be rejected, since there can be no division of the line if $x=0$ or a.

Hence, differentiating again, and substituting $\dfrac{am}{m+n}$ in the second differential coefficient, we have

$$\frac{d^2u}{dx^2} = -(m+n). \quad \text{which indicates a maximum,}$$

$$\therefore x=\frac{am}{m+n} \text{ and } a-x=\frac{an}{m+n} \text{ are the parts.}$$

(5.) If $u=\sin^3 x \cos x$, show that u is a maximum when $x=60°$.

$$\frac{du}{dx} = -\sin^3 x \sin x + \cos x\, 3 \sin^2 x \cos x$$

$$= 3 \sin^2 x \cos^2 x - \sin^4 x = 0,$$

$$\therefore 3 \sin^2 x \cos^2 x = \sin^4 x, \qquad 3 \cos^2 x = \sin^2 x = 1 - \cos^2 x,$$

$$\therefore 4 \cos^2 x = 1, \qquad \cos x = \frac{1}{2}, \qquad \therefore x = 60°,$$

$$\frac{d^2u}{dx^2} = 3 \sin^2 x . 2 \cos x (-\sin x) + 3 \cos^2 x . 2 \sin x \cos x$$

$$-4 \sin^3 x \cos x = -6 \sin^3 x \cos x + 6 \sin x \cos^3 x$$

$$-4 \sin^3 x \cos x = -10 \sin^3 x \cos x + 6 \sin x \cos^3 x.$$

$$\text{Now } \sin x = \frac{\sqrt{3}}{2}, \quad \therefore \sin^3 x = \frac{3\sqrt{3}}{8},$$

$$\therefore \frac{d^2u}{dx^2} = -\frac{30\sqrt{3}}{8} \cdot \frac{1}{2} + \frac{6\sqrt{3}}{2} \cdot \frac{1}{8} = -\frac{24}{16}\sqrt{3}, \text{ a negative result,}$$

$$\therefore u = \frac{3\sqrt{3}}{8} \cdot \frac{1}{2} = \frac{3}{16}\sqrt{3}, \text{ a maximum.}$$

(6.) Divide a number n into two such factors that the sum of their squares shall be the smallest possible.

Let x be one factor, $\dfrac{n}{x}$ the other; then

$$u = x^2 + \frac{n^2}{x^2}, \qquad \frac{du}{dx} = 2x - \frac{2n^2}{x^3} = 0,$$

$$\therefore x = \frac{n^2}{x^3}, \qquad x^4 = n^2, \qquad x = \sqrt{n},$$

$$\frac{d^2u}{dx^2} = 2 + \frac{6n^2}{x^4} = 2 + \frac{6n^2}{n^2} = 2 + 6 = +8, \text{ a positive result,}$$

$\therefore u$ is a minimum. Hence the sum of the squares will be the smallest possible when the factors are equal, each being the square root of the given number.

(7.) Into how many equal parts must a number n be divided that their continued product may be a maximum?

Let there be x equal parts, then

$\dfrac{n}{x}$ is the magnitude of each, and

$$u = \left(\frac{n}{x}\right)^x \text{ is their continued product,}$$

$$\log u = x \log\left(\frac{n}{x}\right) = x\,(\log n - \log x),$$

$$\frac{du}{dx}\frac{1}{u} = x \cdot \left(-\frac{1}{x}\right) + \log n - \log x = -1 + \log n - \log x,$$

$$\frac{du}{dx} = u\{-1 + \log n - \log x\} = 0,$$

$$\therefore \log x = \log n - 1 = \log n - \log e = \log\left(\frac{n}{e}\right), \quad \therefore x = \frac{n}{e}.$$

$$\frac{d^2u}{dx^2} = u\left(-\frac{1}{x}\right) + (-1 + \log n - \log x)\frac{du}{dx}$$

$$= \left(\frac{n}{x}\right)^x\left(-\frac{1}{x}\right) + 0 = e^{\frac{n}{e}}\left(-\frac{e}{n}\right), \text{ a negative result,}$$

$$\therefore u = \left(\frac{n}{x}\right)^x = e^{\frac{n}{e}}, \text{ a maximum.}$$

(8.) Show that $\dfrac{\sin x}{1+\tan x}$ is a maximum when $x=45°$.

$$\frac{du}{dx}=\frac{(1+\tan x)\cos x-\sin x\,(1+\tan^2 x)}{(1+\tan x)^2}$$

$$=\frac{\cos x+\sin x-\sin x-\sin x\tan^2 x}{(1+\tan x)^2}$$

$$=\frac{\cos x-\sin x\tan^2 x}{(1+\tan x)^2}=0,$$

$\therefore \sin x\tan^2 x=\cos x,\quad \dfrac{\sin x}{\cos x}\cdot\tan^2 x=1,\ \tan^3 x=1,\ \therefore x=45°.$

$$\frac{d^2u}{dx^2}=-\frac{3}{4}\sqrt{2},\ \text{a negative result,}$$

$$\therefore u=\frac{\sin x}{1+\tan x}=\frac{1}{4}\sqrt{2},\ \text{a maximum.}$$

(9.) If a be the hypothenuse of a right-angled triangle, find the length of the other sides when the area is a maximum.

Let x be one of the other sides, then

$\sqrt{a^2-x^2}$ is the remaining side.

And area $=\dfrac{1}{2}x\sqrt{a^2-x^2}.$

Now, rejecting the constant $\dfrac{1}{2}$, we may take

$$u=x^2\,(a^2-x^2)=a^2x^2-x^4,$$

$$\frac{du}{dx}=2a^2x-4x^3=2x(a^2-2x^2)=0,\ \therefore x=0,\ x=\frac{a}{\sqrt{2}},$$

$$\frac{d^2u}{dx^2}=2a^2-12x^2=2a^2-6a^2=-4a^2,\ \text{a negative result,}$$

$\therefore u$ is a maximum, and the area is a maximum when the two sides are each $=\dfrac{a}{\sqrt{2}}.$

(10.) What fraction exceeds its nth power by the greatest number possible ?

Let x be the fraction, then $u = x - x^n$, $\dfrac{du}{dx} = 1 - nx^{n-1} = 0$,

$$\therefore nx^{n-1} = 1, \qquad x = \frac{1}{n-\sqrt[1]{n}},$$

$$\frac{d^2u}{dx^2} = -n(n-1)x^{n-2} = -n(n-1) \cdot \frac{1}{n^{\frac{n-2}{n-1}}}, \text{ which is negative,}$$

$\therefore u$ is a maximum.

$$\text{Ans. } \frac{1}{n-\sqrt[1]{n}}.$$

(11.) Within an angle BAC a point P is given, through which it is required to draw a straight line so that the triangle cut off by it shall be the smallest possible.

Let $PN = a$, $AN = b$, $AD = x$, then $ND = x - b$, $ND : PN :: AD : AE$ or

$$x - b : a :: x : AE, \qquad \therefore AE = \frac{ax}{x-b}.$$

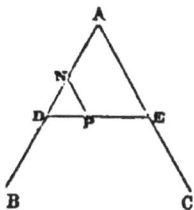

Now area $\triangle DAE = \dfrac{1}{2} AD \cdot AE \sin A = \dfrac{1}{2} x \cdot \dfrac{ax}{x-b} \sin A,$

$$\therefore u = \frac{x^2}{x-b}, \quad \frac{du}{dx} = \frac{(x-b)\,2x - x^2}{(x-b)^2} = \frac{x^2 - 2bx}{(x-b)^2} = \frac{x(x-2b)}{(x-b)^2} = 0,$$

$$\therefore x = 2b,$$

$$\frac{d^2u}{dx^2} = \frac{(x-b)^2 \cdot (2x - 2b) - (x^2 - 2bx) \cdot 2(x-b)}{(x-b)^4}$$

$$= \frac{2(x-b)^2 - 2(x^2 - 2bx)}{(x-b)^3} = \frac{2b^2}{(x-b)^3} = \frac{2b^2}{b^3} = +\frac{2}{b},$$

a positive result, $\therefore$ the area is a minimum.

Since $AD = 2AN$, $\therefore DE = 2DP$, $\therefore$ the line must be so drawn as to be bisected by the given point P.

(12.) From two points A, B, to draw two straight lines to a point P in a given line ON, so that $AP+BP$ shall be a minimum.

Let O be the origin of co-ordinates, and the given line the axis of x.

Let $OP=x$, and let the co-ordinates of A be a, b, and those of B be $a_{,}$, $b_{,}$. Then

$$AP = \sqrt{AM^2+PM^2} = \sqrt{b^2+(x-a)^2},$$

$$BP = \sqrt{BN^2+PN^2} = \sqrt{b_{,}^2+(a_{,}-x)^2},$$

$$\therefore u = AP+BP = \sqrt{b^2+(x-a)^2} + \sqrt{b_{,}^2+(a_{,}-x)^2}, \text{ a minimum,}$$

$$\frac{du}{dx} = \frac{x-a}{\sqrt{b^2+(x-a)^2}} - \frac{a_{,}-x}{\sqrt{b_{,}^2+(a_{,}-x)^2}} = 0,$$

$$\therefore \frac{x-a}{\sqrt{b^2+(x-a)^2}} = \frac{a_{,}-x}{\sqrt{b_{,}^2+(a_{,}-x)^2}}, \text{ or}$$

$$\frac{MP}{AP} = \frac{NP}{BP}, \quad \therefore \angle APM = BPN.$$

(13.) If the length of an arc of a circle be $2a$, find the angle it must subtend at the centre so that the corresponding segment may be a maximum or minimum.

Draw CD bisecting the arc, and let x be the radius, then $\dfrac{a}{x} = \angle ACD$.

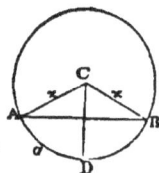

Now area segment $ADB = \text{sector } ACB - \triangle ACB$

$$= \frac{1}{2} \text{ rad} \times \text{arc} - \frac{1}{2} x^2 \sin ACB$$

$$= ax - \frac{1}{2} x^2 . 2 \sin \frac{ACB}{2} \cos \frac{ACB}{2}$$

$$\therefore u = ax - x^2 \sin \frac{a}{x} \cos \frac{a}{x},$$

$$\frac{du}{dx} = a - x^2_? \sin \frac{a}{x} \left(-\sin \frac{a}{x} \right) \left(-\frac{a}{x^2} \right) - x^2 \cos \frac{a}{x} \cos \frac{a}{x} \left(-\frac{a}{x^2} \right)$$

$$- 2x \sin \frac{a}{x} \cos \frac{a}{x}$$

$$= a - a \sin^2 \frac{a}{x} + a \cos^2 \frac{a}{x} - x 2 \sin \frac{a}{x} \cos \frac{a}{x}$$

$$= a - a + 2a \cos^2 \frac{a}{x} - x 2 \sin \frac{a}{x} \cos \frac{a}{x}$$

$$= 2 \cos \frac{a}{x} \left(a \cos \frac{a}{x} - x \sin \frac{a}{x} \right) = 0.$$

Take $\cos \frac{a}{x} = 0$, $\therefore \frac{a}{x} = \frac{\pi}{2}$, $x = \frac{2a}{\pi}$, and the segment is a $\frac{1}{2} \odot$

$$= \text{maximum.}$$

Take $a \cos \frac{a}{x} = x \sin \frac{a}{x}$, $\qquad \therefore \dfrac{\sin \frac{a}{x}}{\cos \frac{a}{x}} = \tan \frac{a}{x} = \frac{a}{x}$,

$$\therefore \frac{a}{x} = 0, \qquad \therefore x = \infty \quad \text{and } u = \text{minimum.}$$

> (14.) Within a given circle to inscribe the greatest isosceles triangle.

Let radius $OA = a$, $AB = AC = x$, $BC = 2y$,

$BD = y$. Then $\triangle = \dfrac{AB \cdot AC \cdot BC}{4 AO} = \dfrac{x^2 y}{2a}$,

Also $\triangle = \dfrac{BC \cdot AD}{2} = \dfrac{BC \sqrt{AB^2 - BD^2}}{2} = y \sqrt{x^2 - y^2}$,

$$\therefore \frac{x^2 y}{2a} = y \sqrt{x^2 - y^2}, \qquad x^2 = 2a \sqrt{x^2 - y^2},$$

$$x^4 = 4a^2 x^2 - 4a^2 y^2, \quad 4a^2 y^2 = 4a^2 x^2 - x^4, \quad 2ay = x \sqrt{4a^2 - x^2},$$

$$\therefore y = \frac{1}{2a} x \sqrt{4a^2 - x^2},$$

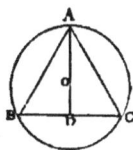

Now $\Delta = \dfrac{1}{2a} x^2 y = \dfrac{1}{2a} x^2 \cdot \dfrac{1}{2a} x \sqrt{4a^2 - x^2}$, a maximum.

Put $u = x^6(4a^2 - x^2) = 4a^2 x^6 - x^8$,

$\dfrac{du}{dx} = 24 a^2 x^5 - 8 x^7 = 0, \quad \therefore 8 x^7 = 24 a^2 x^5,$

$$x^2 = 3a^2, \quad \therefore x = a\sqrt{3},$$

$BC = 2y = \dfrac{1}{a} a\sqrt{3} \sqrt{4a^2 - 3a^2} = a\sqrt{3}$, and Δ is equilateral.

(15.) Of all equiangular and isoperimetrical parallelograms, show that the equilateral has the greatest area.

The perimeters of the figures being all equal, the perimeter of each may be considered as one line, and the proposition then resolves itself into the following. " To divide a given straight line into two such parts that the rectangle contained by those parts shall be the greatest possible."

Let a be the line, x one part, then $a - x$ is the other, $x(a - x)$ is the rectangle, and $u = ax - x^2$, a maximum.

$$\frac{du}{dx} = a - 2x = 0, \qquad \therefore x = \frac{a}{2}.$$

$\therefore$ the line must be divided into two *equal* parts, and the parallelogram will be equilateral.

(16.) Of all triangles on the same base and having equal vertical angles the isosceles has the greatest perimeter.

Let a be the base, α the vertical angle, x and y the two sides, then $u = a + x + y = a$ maximum.

$$\frac{du}{dx} = 1 + \frac{dy}{dx} = 0, \quad \therefore \frac{dy}{dx} = -1, \quad 2xy \cos\alpha = x^2 + y^2 - a^2,$$

$$2\cos\alpha \cdot x \frac{dy}{dx} + 2y \cos\alpha = 2x + 2y\frac{dy}{dx}; \quad -x\cos\alpha + y\cos\alpha = x - y,$$

$$\therefore -(x-y)\cos\alpha = x - y, \quad \therefore x - y = 0,$$

$$\therefore x = y, \text{ and the } \Delta \text{ is isosceles.}$$

> (17.) The segment of a circle being given, it is required
to inscribe the greatest possible rectangle in it.

Let BAD be the segment, radius $= a$,
$AM=x$, draw AC through the centre perpendicular to PM or BD. Let $AC=b$.

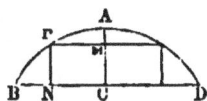

Then $PM^2=(2a-x)x$,   Euc. B. iii. p. 35.

$$\therefore PM = \sqrt{2ax-x^2}, \qquad MC=b-x,$$

Area rectangle $=MC \cdot 2\,PM=2\,(b-x)\,\sqrt{2ax-x^2}.$

Put $u=(b-x)^2 \cdot (2ax-x^2)$,

$$\frac{du}{dx}=(b-x)^2(2a-2x)+(2ax-x^2)\cdot 2\,(b-x)\,(-1)=0$$

$$\therefore (b-x)\,(a-x)=2ax-x^2, \qquad ab-ax-bx+x^2=2ax-x^2,$$

$$2x^2-(3a+b)\,x=-ab, \qquad x^2-\frac{3a+b}{2}\,x=-\frac{ab}{2},$$

$$\therefore x=\frac{3a+b\pm \sqrt{9a^2-2ab+b^2}}{4}.$$

> (18.) To cut the greatest parabola from a given right
cone.

Let $BD=a$, $AD=b$, $BC=x$, $CD=a-x$,
Then $\because BNDM$ is a circle, and $MC=NC$,

$$\therefore MC^2=BC\cdot CD, \qquad MC= \sqrt{x(a-x)},$$

$$MN=2\sqrt{ax-x^2}.$$

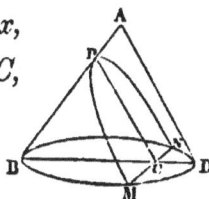

Also $BD : AD :: BC : PC$, $\quad \therefore PC=\dfrac{AD \cdot BC}{BD}=\dfrac{bx}{a}$,

Area parabola $=\dfrac{2}{3}\,PC\cdot MN=\dfrac{2}{3}\,\dfrac{bx}{a}\cdot 2\sqrt{ax-x^2}$, a maximum.

Put $u=x^2(ax-x^2)=ax^3-x^4$,

$$\frac{du}{dx}=3ax^2-4x^3=0, \qquad \therefore 4x^3=3ax^2, \qquad x=\frac{3}{4}\,a.$$

$$\frac{d^2u}{dx^2}=-\frac{9}{4}\,a^2, \text{ which indicates a maximum.}$$

(19.) Within a given parabola to inscribe the greatest parabola, the vertex of the latter being at the bisection of the base of the former.

Let BAC be the given parabola, L its latus rectum.

$$AD=a, \quad BD=b, \quad DN=x, \quad PN=y.$$

$$\text{Area parabola} = \frac{2}{3} \cdot 2\,PN \cdot ND = \frac{2}{3} \cdot 2\,yx.$$

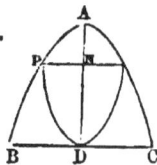

Now ∵ the square of any ordinate to the axis = the rectangle under the latus rectum and abscissa,

$$\therefore y^2 = L \cdot AN = L\,(a-x), \quad b^2 = L \cdot AD = L \cdot a,$$

$$\therefore \frac{y^2}{b^2} = \frac{a-x}{a}, \qquad y = \frac{b}{\sqrt{a}}\sqrt{a-x},$$

$$\therefore \text{area parabola} = \frac{4}{3}\frac{b}{\sqrt{a}}x\sqrt{a-x}.$$

$$\text{Put } u = x^2\,(a-x) = ax^2 - x^3,$$

$$\frac{du}{dx} = 2ax - 3x^2 = 0, \quad \therefore 3x^2 = 2ax, \qquad x = \frac{2}{3}a.$$

> (20.) Inscribe the greatest cylinder within a given right cone.

Let ABC be the cone, $AD=a, \quad BD=b, \quad DN=x, \quad PN=y,$ $AN=a-x.$

$$\text{Volume of cylinder} = \frac{\pi}{4} \cdot (2PN)^2 .\, ND = \pi y^2 x.$$

$$AD : BD :: AN : PN, \quad \therefore PN = \frac{BD}{AD} \cdot AN, \text{ or }$$

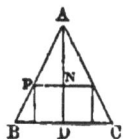

$$y = \frac{b}{a}(a-x), \quad \therefore \text{cylinder} = \pi \cdot \frac{b^2}{a^2}(a-x)^2 x.$$

$$\text{Put } u = (a-x)^2 x = a^2 x - 2ax^2 + x^3,$$

$$\frac{du}{dx} = a^2 - 4ax + 3x^2 = 0, \quad \therefore 3x^2 - 4ax = -a^2, \quad x = a \text{ or } \frac{a}{3}.$$

$$y = \frac{b}{a}\left(a - \frac{a}{3}\right) = \frac{2}{3}b, \quad \text{cylinder} = \pi \cdot \frac{4b^2}{9} \cdot \frac{a}{3} = \frac{4\pi a b^2}{27}.$$

> (21.) If the volume of a cylinder be a, find its form when its surface is the least possible.

Let $AB = x$, $BC = y$.

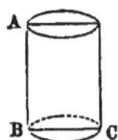

Surface = convex surface + 2 area of base

$$= BC \cdot \pi \cdot AB + 2 BC^2 \cdot \frac{\pi}{4} = \pi xy + \frac{\pi}{2} y^2.$$

Volume $= a = BC^2 \cdot \frac{\pi}{4} \cdot AB = \frac{\pi}{4} xy^2$, $\quad \therefore x = \frac{4a}{\pi y^2}$

Hence $u = \pi \cdot \frac{4a}{\pi y^2} y + \frac{\pi}{2} y^2 = \frac{4a}{y} + \frac{\pi}{2} y^2$,

$$\frac{du}{dy} = -\frac{4a}{y^2} + \pi y = 0, \qquad \therefore y^3 = \frac{4a}{\pi},$$

$$x^3 = \frac{64 a^3}{\pi^3 y^6} = \frac{64 a^3}{\pi^3 \cdot \frac{16 a^2}{\pi^2}} = \frac{4a}{\pi}, \qquad \therefore x = y,$$

or altitude = diameter of base.

$$\frac{d^2 u}{dy^2} = \frac{8 ay}{y^4} + \pi = \frac{8a}{y^3} + \pi = +3\pi, \text{ a positive result,}$$

$\therefore$ the surface is a minimum.

> (22.) The latitude of a place and two circles parallel to the horizon being given ; to determine the declination of a heavenly body, whose apparent time of passage from one circle to the other shall be a minimum.

Let P be the pole, Z the zenith, S, S_1 the positions of the heavenly body on the parallel circles, the polar distances PS, PS_1 being equal,

$\angle ZPS = P$, $ZPS_1 = P_1$, polar distance PS or $PS_1 = x$, arc $ZS = a$, $ZS_1 = a_1$, latitude $= l$, declination $= \delta$; then

$\because$ the passage along the arc SS_1 is the shortest possible,

$\therefore$ the angle $SPS_1 =$ a minimum,

$$\therefore \frac{dSPS_,}{dx} = \frac{d(P_, - P)}{dx} = 0, \qquad \therefore \frac{dP_,}{dx} = \frac{dP}{dx},$$

But $\dfrac{dP}{dx} = -\dfrac{\cot S}{\sin x},$ $\qquad \dfrac{dP_,}{dx} = -\dfrac{\cot S_,}{\sin x},$

$$\therefore \frac{\cot S}{\sin x} = \frac{\cot S_,}{\sin x}, \qquad \therefore S = S_,.$$

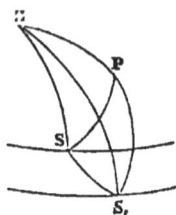

Again $\cos S = \dfrac{\sin l - \cos a \cos x}{\sin a \, \sin x},$ $\qquad \cos S_, = \dfrac{\sin l - \cos a_, \cos x}{\sin a_, \sin x},$

$$\therefore \frac{\sin l - \cos a \cos x}{\sin a} = \frac{\sin l - \cos a_, \cos x}{\sin a_,},$$

$$\cos x = \frac{\cos \dfrac{1}{2}(a_, + a)}{\cos \dfrac{1}{2}(a_, - a)} \cdot \sin l.$$

And $\because$ the declination is the complement of the polar

distance, $\qquad \therefore \sin \delta = \dfrac{\cos \dfrac{1}{2}(a_, + a)}{\cos \dfrac{1}{2}(a_, - a)} \cdot \sin l.$

Cor. If $a = \dfrac{\pi}{2}$, and $a_, = \dfrac{\pi}{2} + 2d$, this expression becomes $\sin \delta = -\tan d \sin l$; and if the heavenly body be the sun, and $2d = 18°$ nearly $=$ his depression below the horizon when twilight begins in the morning or ends in the evening, we are enabled to determine the time of shortest twilight by means of the analogy rad : sin lat :: tan 9° : $-\sin \delta$, where the negative sign indicates that, if the latitude be north, the declination will be south, and *vice versa*.

> (23.) The centres of two spheres (radii r_1, r_2) are at the extremities of a straight line $2a$, on which a circle is described.

Find a point in the circumference from which the greatest portion of spherical surface is visible.

Let x and y be the distances of the point from the centres of the two spheres; draw tangents EA, EB, ED, EF; join AB, DF.

Then, of the sphere C the portion visible is the convex surface of the segment $AHBS$, whose area = height HS × circumference of the sphere.

Now $x : r_1 :: r_1 : CS$, $\therefore CS = \dfrac{r_1{}^2}{x}$, $\therefore$ height of segment

$$= r_1 - \frac{r_1{}^2}{x} = HS, \qquad \text{circumference of sphere} = 2\pi r_1,$$

$$\therefore 2\pi r_1\left(r_1 - \frac{r_1{}^2}{x}\right) = \text{visible portion of sphere } C; \text{ and similarly}$$

$$2\pi r_2\left(r_2 - \frac{r_2{}^2}{y}\right) = \text{visible portion of sphere } c.$$

Hence $2\pi\left\{r_1\left(r_1 - \dfrac{r_1{}^2}{x}\right) + r_2\left(r_2 - \dfrac{r_2{}^2}{y}\right)\right\} = $ whole visible surface.

Put $u = r_1{}^2 - \dfrac{r_1{}^3}{x} + r_2{}^2 - \dfrac{r_2{}^3}{y}$,   then

$$\frac{du}{dx} = \frac{r_1{}^3}{x^2} + \frac{r_2{}^3\frac{dy}{dx}}{y^2} = 0, \qquad \therefore \frac{r_1{}^3}{x^2} = -\frac{r_2{}^3}{y^2}\frac{dy}{dx},$$

But $y = \sqrt{4a^2 - x^2}$, $\therefore \dfrac{dy}{dx} = -\dfrac{x}{\sqrt{4a^2 - x^2}}$,

Hence $\dfrac{r_1{}^3}{x^2} = \dfrac{r_2{}^3}{4a^2 - x^2} \cdot \dfrac{x}{\sqrt{4a^2 - x^2}} = \dfrac{r_2{}^3 x}{(4a^2 - x^2)^{\frac{3}{2}}}$,

$\therefore r_1{}^3(4a^2 - x^2)^{\frac{3}{2}} = r_2{}^3 x^3$, $r_1(4a^2 - x^2)^{\frac{1}{2}} = r_2 x$, $r_1{}^2(4a^2 - x^2) = r_2{}^2 x^2$,

$4a^2 r_1{}^2 - r_1{}^2 x^2 = r_2{}^2 x^2$, $\qquad x^2 = \dfrac{4a^2 r_1{}^2}{r_1{}^2 + r_2{}^2}$, $\qquad x = \dfrac{2ar_1}{\sqrt{r_1{}^2 + r_2{}^2}}$.

(24.) Of all ellipses that can be inscribed in a rhombus whose diagonals are $2m$ and $2n$, show that the greatest is that whose major and minor semi-axes are $\dfrac{m}{\sqrt{2}}$ and $\dfrac{n}{\sqrt{2}}$ respectively.

$ABCD$ the rhombus, $OC=m$, $OB=n$, a and b the semi-axes of the ellipse.

Let $ON=x$, $NP=y$. Then by the properties of the ellipse

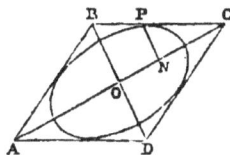

$$OC \cdot ON = a^2, \qquad OB \cdot NP = b^2, \qquad \text{or } m \cdot x = a^2, \qquad n \cdot y = b^2,$$

$$\therefore m^2 x^2 = a^2 \cdot a^2, \qquad n^2 y^2 = b^2 \cdot b^2, \qquad \frac{x^2}{a^2} = \frac{a^2}{m^2}, \qquad \frac{y^2}{b^2} = \frac{b^2}{n^2},$$

$$\therefore \frac{x^2}{a^2} + \frac{y^2}{b^2} = \frac{a^2}{m^2} + \frac{b^2}{n^2} = 1; \quad \ldots \text{ (1)}, \qquad \text{where } a \text{ and } b \text{ alone}$$

must be considered as variables.

But, area ellipse $= \pi a b =$ a maximum.

Rejecting the constant π, and differentiating this and equation (1), we have

$$b + a \frac{db}{da} = 0, \qquad \frac{a}{m^2} + \frac{b}{n^2} \cdot \frac{db}{da} = 0,$$

$$\therefore \frac{a}{m^2} - \frac{b}{n^2} \cdot \frac{b}{a} = 0, \qquad \frac{a^2}{m^2} = \frac{b^2}{n^2}, \qquad \frac{2a^2}{m^2} = \frac{2b^2}{n^2} = 1,$$

$$\frac{a}{m} = \frac{1}{\sqrt{2}}, \qquad \frac{b}{n} = \frac{1}{\sqrt{2}}, \qquad \therefore a = \frac{m}{\sqrt{2}}, \qquad b = \frac{n}{\sqrt{2}}.$$

(25.) If $u = x^4 - 8x^3 + 22x^2 - 24x + 12$, find the values of x which render u a maximum or a minimum.

Ans. When $x=3$, u is a minimum,
$x=2$, u is a maximum,
$x=1$, u is a minimum.

(26.) Find when $x^3 - 6x^2 + 9x + 10$ is a maximum, and when it is a minimum.

When $x=3$, u is a minimum,
$x=1$, u is a maximum.

(27.) Find the maxima and minima values of the function $u = 3a^2x^3 - b^4x + c^5$.

$$\text{When } x = \frac{b^2}{3a}, \; u \text{ is a minimum.}$$

$$x = -\frac{b^2}{3a}, \; u \text{ is a maximum.}$$

(28.) $u = \dfrac{a^2x}{(a-x)^2}$; ascertain when u is a maximum and when a minimum.

$$\text{When } x = -a, \; u = -\frac{a}{4}, \text{ a minimum,}$$

$$x = +a, \; u = \infty, \text{ a maximum.}$$

(29.) $u = x^{\frac{1}{x}}$; find when u is a maximum.

$$x = e = 2 \cdot 71828 \text{ &c.}$$

(30.) $u = \dfrac{(x+3)^3}{(x+2)^2}$; determine when u is a maximum and when a minimum.

$$x = -2, \; u = \infty, \text{ a maximum,}$$

$$x = 0, \; u = 6\tfrac{3}{4}, \text{ a minimum.}$$

(31.) $u = x + \sqrt{a^2 - 2bx + b^2}$; when is u a maximum?

$$\text{When } x = \frac{a^2}{2b}, \; u = \frac{a^2}{2b} + b, \text{ a maximum.}$$

(32.) $u = \dfrac{ab}{x\sqrt{a^2 + b^2 - x^2}}$; show that u is a minimum when $x = \sqrt{\dfrac{a^2 + b^2}{2}}$.

(33.) $u = \sec x + \operatorname{cosec} x$; show that u is a minimum when $x = \dfrac{\pi}{4}$.

(34.) In a given triangle to inscribe the greatest parallelogram.

Ans. Side of parallelogram = $\frac{1}{2}$ side of triangle.

(35.) A column a feet high has a statue on the top of it, the height from the ground to the top of the statue is b feet ;

$(h-a)(x'-a) = a$

find a point in the horizontal plane at which the statue sub-
tends the greatest angle.   *Ans.* $\sqrt{ab}$ feet from the base.

> (36.) Show that the difference between the sine and
versed sine is a maximum when the arc is 45°.

(37.) Let AC and BD be parallel, and join
AD; it is required to draw from C a straight
line so that the triangles EOD, AOC together
shall be a minimum.

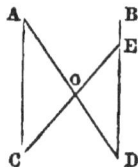

Let $AC=a$, $AD=b$, $AO=x$; then $x=\sqrt{b}$.

(38.) The base and vertical angle of a triangle being given,
show that when it is isosceles its area is a maximum.

(39.) A farmer has a field of triangular form, which he
wishes to divide into two equal parts by a fence; find the
points in the sides of the field from which he must draw the
line, for his fence to be the least possible expense to him.

Ans. If a, b, c be the sides, the distance of each point

from the angle C is $\sqrt{\dfrac{ab}{2}}$, and the length

of the fence is $\sqrt{\dfrac{(c-a+b)(c+a-b)}{2}}$.

(40.) If the greatest rectangle be inscribed in an ellipse,
the greatest ellipse in that rectangle, again the greatest rectangle in that ellipse, and so on continually; show that the
sum of all the inscribed rectangles is equal to the area of any
parallelogram circumscribing the given ellipse.

(41.) Prove that the greatest area that can be contained
by four straight lines is that of a quadrilateral inscribed in a
circle.

(42.) Inscribe the greatest ellipse in a given isosceles
triangle. *Ans.* Major axis $=\dfrac{2}{3}$ altitude of triangle.

(43.) A tree, in the form of a frustrum of a cone, is n feet long, and its greater and less diameters are a and b feet respectively; show that the greatest square beam that can be cut out of it is $\dfrac{na}{3(a-b)}$ feet long.

(44.) Describe the least isosceles triangle about a given circle.   The triangle is equilateral.

(45.) To inscribe the greatest right cone in a given sphere, whose radius is r.

Distance of base of cone from centre of sphere $=\dfrac{r}{3}$.

(46.) If the polar diameter of the earth be to the equatorial diameter as $229 : 230$; show that the greatest angle made by a body falling to the earth, with a perpendicular to the surface, is $14' 58''$, and that the latitude is $45° 7' 29''$. See fig. ex. 9. page 84.

(47.) In a parabolic curve, whose vertex is A, and focus S, find a point P, such that the ratio $AP : SP$ shall be a maximum. $AP : SP :: 2 : \sqrt{3}$.

(48.) Inscribe the greatest parabola in a given isosceles triangle.   Altitude of parabola $=\dfrac{3}{4}$ altitude of triangle.

(49.) If in a circle, whose radius is r, a right-angled triangle be inscribed; show that, when a maximum circle is inscribed in the triangle, the area of the triangle is r^2.

(50.) Inscribe the greatest cylinder in a given prolate spheroid.

(51.) Required the maximum and minimum values of u in the equation $u^3 - a^2 x + x^3 = 0$.

(52.) $u = \dfrac{e^{\cos x}}{\cos^n x}$; find the maximum and minimum values of u.

(53.) Show that the greatest paraboloid that can be inscribed in a given right cone is $\frac{2}{3}$ of the height of that cone.

(54.) $u = x^{1-\log x}$; show that when u is a maximum, $\log x = \frac{1}{2}$.

(55.) Find that sphere which, being put into a conical vessel of given dimensions, will displace the greatest possible quantity of fluid.

(56.) Two circles of given radii intersect each other; find the longest straight line which can be drawn through either point of intersection, and terminated by the circumferences.

(57.) If a tangent to a great circle of a sphere measure $5\frac{1}{2}$, and a perpendicular to a tangent meeting the great circle measure 4 feet; show that the volume of the sphere is to the volume of its greatest inscribed semispheroid as 27 : 16.

(58.) Find what values of x make $(x-2)(x+3)(5-x)$ a maximum or minimum, and distinguish the one from the other.

(59.) Inscribe the greatest cone in a given hemisphere ABC, the vertex of the cone being at A.

For other examples and solutions see chap. xi.

IMPLICIT FUNCTIONS OF TWO VARIABLES.

If $u = f(x, y)$, u being an implicit function of the two variables x and y, by putting $\frac{du}{dx} = 0$, we shall find the values of x which render y a maximum or minimum.

By substituting the particular value of x in $\left(\frac{d^2u}{dx^2} \div \frac{du}{dy}\right)$, if the result be positive, y will be a maximum; if negative, a minimum.

Ex. (1.) Let $u=x^3-3a^2x+y^3=0$; determine the maximum and minimum values of y.

Differentiate with respect to x, considering y constant.

$$\frac{du}{dx}=3x^2-3a^2=0, \quad \therefore x^2=a^2, \quad x=+a, \quad x=-a.$$

$$\frac{d^2u}{dx^2}=6x. \qquad \text{Differentiate the given function with}$$

respect to y, considering x constant.

$$\frac{du}{dy}=3y^2. \qquad \text{Substitute the values of } x \text{ in } u.$$

$$a^3-3a^3+y^3=0, \quad \therefore y^3=2a^3, \quad y=a\sqrt[3]{2},$$

$$-a^3+3a^3+y^3=0, \quad \therefore y^3=-2a^3, \quad y=-a\sqrt[3]{2}.$$

$$\therefore \frac{d^2u}{dx^2} \div \frac{dy}{dx}=\frac{6x}{3y^2}=\frac{6a}{3a^2.2^{\frac{2}{3}}}=+\frac{\sqrt[3]{2}}{a}, \quad \text{a positive re-}$$

sult, $\therefore y=a\sqrt[3]{2}$ is a maximum.

$$\frac{d^2u}{dx^2} \div \frac{dy}{dx}=\frac{6x}{3y^2}=\frac{-6a}{3a^2.2^{\frac{2}{3}}}=-\frac{\sqrt[3]{2}}{a}, \quad \text{a negative re-}$$

sult, $\therefore y=-a\sqrt[3]{2}$ is a minimum.

(2.) $u=x^3-3axy+y^3=0$; show that when $x=0$, $y=0$, a minimum; and when $x=a\sqrt[3]{2}$, $y=a\sqrt[3]{4}$, a maximum.

(3.) $4xy-y^4-x^4=2$; show that when $x=+1$ or -1, $y=+1$ or -1, neither being a maximum or minimum.

(4.) $y^2-3=-2x(xy+2)$; show that when $x=1$, $y=-1$, neither a maximum nor a minimum; but when $x=-\frac{1}{2}$, $y=2$, a maximum.

CHAPTER VIII.

FUNCTIONS OF TWO OR MORE VARIABLES.

If $u=f(x, y)$, x and y being two variables independent of each other, then

$$\frac{d^2u}{dy\,dx}=\frac{d^2u}{dx\,dy}, \qquad \frac{d^3u}{dy^2\,dx}=\frac{d^3u}{dx\,dy^2}, \qquad \frac{d^3u}{dy\,dx^2}=\frac{d^3u}{dx^2\,dy},$$

and generally $\dfrac{d^{n+r}u}{dy^r dx^n}=\dfrac{d^{n+r}u}{dx^n dy^r}.$

In a function of any number of variables, the order of differentiation is indifferent.

The total differential of two variables is equal to the sum of the partial differentials ; or if $u=f(x, y)$,

$$du= \left(\frac{du}{dx}\right) dx + \left(\frac{du}{dy}\right) dy.$$

$$d^nu=\frac{d^nu}{dx^n} dx^n + n \frac{d^nu}{dx^{n-1}dy} dx^{n-1}dy$$

$$+\frac{n(n-1)}{1\cdot 2} \frac{d^nu}{dx^{n-2}dy^2} dx^{n-2}dy^2 + \&c.$$

Ex. (1.) Let $u=x^3y^2$; find du, and $\dfrac{d^2u}{dx\,dy}$.

To find the partial differential coefficient $\left(\dfrac{du}{dx}\right)$, consider y constant, and differentiate with respect to x ; and to find $\left(\dfrac{du}{dy}\right)$, consider x constant, and differentiate with respect to y.

To find $\dfrac{d^2u}{dy\,dx}$ or $\dfrac{d^2u}{dx\,dy}$, differentiate $\left(\dfrac{du}{dx}\right)$ considering x constant, or differentiate $\left(\dfrac{du}{dy}\right)$ considering y constant.

$$u=x^3y^2,$$

$$\left(\frac{du}{dx}\right)=3\,x^2y^2, \qquad \left(\frac{du}{dy}\right)=2\,yx^3,$$

$$du=\left(\frac{du}{dx}\right)dx+\left(\frac{du}{dy}\right)dy=3\,x^2y^2\,dx+2\,yx^3\,dy$$
$$=x^2y\,(3\,y\,dx+2\,x\,dy).$$

$$\frac{d^2u}{dy\,dx}=3\times2\,yx^2=6\,x^2y=\frac{d^2u}{dx\,dy}.$$

(2.) $u=\dfrac{x^2+y^2}{x^2-y^2}$; find $\dfrac{d^2u}{dy\,dx}$.

$$\left(\frac{du}{dx}\right)=\frac{(x^2-y^2)\cdot2\,x-(x^2+y^2)\cdot2\,x}{(x^2-y^2)^2}=-\frac{4\,xy^2}{(x^2-y^2)^2},$$

$$\frac{d^2u}{dy\,dx}=-\frac{(x^2-y^2)^2\cdot8\,xy-4\,xy^2\cdot2\,(x^2-y^2)\,(-2\,y)}{(x^2-y^2)^4}$$

$$=-\frac{8\,xy\,(x^2-y^2)+16\,xy^3}{(x^2-y^2)^3}=-\frac{8\,x^3y+8\,xy^3}{(x^2-y^2)^3}$$

$$=-8\,xy\,\frac{x^2+y^2}{(x^2-y^2)^3}=\frac{d^2u}{dx\,dy}.$$

(3.) $u=\sin^{-1}\dfrac{x}{y}$; find du, and $\dfrac{d^2u}{dy\,dx}$,

$$\sin u=\frac{1}{y}\,x, \qquad \cos u\,\frac{du}{dx}=\frac{1}{y},$$

$$\therefore\left(\frac{du}{dx}\right)=\frac{1}{y\cos u}=\frac{1}{y\sqrt{1-\sin^2u}}=\frac{1}{y\sqrt{1-\dfrac{x^2}{y^2}}}$$

$$=\frac{1}{y\sqrt{\dfrac{y^2-x^2}{y^2}}}=\frac{1}{\sqrt{y^2-x^2}}.$$

Again $\sin u = \dfrac{x}{y}$, consider x constant.

$$\cos u \cdot \frac{du}{dy} = -\frac{x}{y^2},$$

$$\therefore \left(\frac{du}{dy}\right) = -\frac{x}{y^2\cos u} = -\frac{x}{y^2\sqrt{1-\sin^2 u}} = -\frac{x}{y\sqrt{y^2-x^2}}.$$

Hence $du = \left(\dfrac{du}{dx}\right)dx + \left(\dfrac{du}{dy}\right)dy$

$$= \frac{1}{\sqrt{y^2-x^2}}dx - \frac{x}{y\sqrt{y^2-x^2}}dy = \frac{y\,dx - x\,dy}{y\sqrt{y^2-x^2}}.$$

$$\frac{d^2u}{dy\,dx} = -\frac{\dfrac{y}{\sqrt{y^2-x^2}}}{y^2-x^2} = \frac{-y}{(y^2-x^2)^{\frac{3}{2}}} = \frac{d^2u}{dx\,dy}.$$

(4.) $u = \dfrac{x^2 y}{a^2-z^2}$; find du, and show that

$$\frac{d^2u}{dx\,dy} = \frac{2x}{a^2-z^2} = \frac{d^2u}{dy\,dx}, \qquad \frac{d^2u}{dx\,dz} = \frac{4xyz}{(a^2-z^2)^2} = \frac{d^2u}{dz\,dx},$$

$$\frac{d^2u}{dy\,dz} = \frac{2x^2z}{(a^2-z^2)^2} = \frac{d^2u}{dz\,dx}, \qquad \text{and}$$

$$\frac{d^3u}{dx\,dy\,dz} = \frac{4xz}{(a^2-z^2)^2} = \frac{d^3u}{dz\,dy\,dx} = \frac{d^3u}{dy\,dx\,dz}.$$

First differentiate considering y, z constant; then considering x, z constant; and lastly considering x, y constant.

$$\frac{du}{dx} = \frac{2xy}{a^2-z^2}, \qquad \frac{du}{dy} = \frac{x^2}{a^2-z^2}, \qquad \frac{du}{dz} = \frac{2x^2yz}{(a^2-z^2)^2},$$

$$du = \left(\frac{du}{dx}\right)dx + \left(\frac{du}{dy}\right)dy + \left(\frac{du}{dz}\right)dz$$

$$= \frac{2xy}{a^2-z^2}dx + \frac{x^2}{a^2-z^2}dy + \frac{2x^2yz}{(a^2-z^2)^2}dz.$$

Now $\dfrac{du}{dx} = \dfrac{2\,xy}{a^2 - z^2}$.   Consider x, z constant, and differentiate.

$$\frac{d^2u}{dx\,dy} = \frac{2\,x}{a^2 - z^2}.$$

$\dfrac{du}{dy} = \dfrac{x^2}{a^2 - z^2}$.   Consider z constant, and differentiate.

$\dfrac{d^2u}{dy\,dx} = \dfrac{2\,x}{a^2 - z^2}$.   Hence $\dfrac{d^2u}{dx\,dy} = \dfrac{2\,x}{a^2 - z^2} = \dfrac{d^2u}{dy\,dx}$.

Again $\dfrac{du}{dx} = \dfrac{2\,xy}{a^2 - z^2}$.   Consider x, y constant, and differentiate.

$$\frac{d^2u}{dx\,dz} = \frac{-2\,xy \cdot (-2\,z)}{(a^2 - z^2)^2} = \frac{4\,xyz}{(a^2 - z^2)^2}.$$

$\dfrac{du}{dz} = \dfrac{2\,x^2yz}{(a^2 - z^2)^2}$.   Consider y, z constant, and differentiate.

$\dfrac{d^2u}{dz\,dx} = \dfrac{4\,xyz}{(a^2 - z^2)^2}$.   Hence $\dfrac{d^2u}{dx\,dz} = \dfrac{4\,xyz}{(a^2 - z^2)^2} = \dfrac{d^2u}{dz\,dx}$.

Again $\dfrac{du}{dy} = \dfrac{x^2}{a^2 - z^2}$.   Consider x constant, and differentiate.

$$\frac{d^2u}{dy\,dz} = \frac{-x^2 \cdot (-2\,z)}{(a^2 - z^2)^2} = \frac{2\,x^2z}{(a^2 - z^2)^2}.$$

$\dfrac{du}{dz} = \dfrac{2\,x^2yz}{(a^2 - z^2)^2}$.   Consider x, z constant, and differentiate.

$\dfrac{d^2u}{dz\,dy} = \dfrac{2\,x^2z}{(a^2 - z^2)^2}$.   Hence $\dfrac{d^2u}{dy\,dz} = \dfrac{2\,x^2z}{(a^2 - z^2)^2} = \dfrac{d^2u}{dz\,dy}$.

Now $\dfrac{d^2u}{dx\,dy} = \dfrac{2\,x}{a^2 - z^2}$.   Consider x, y constant, and differentiate.

$$\frac{d^3u}{dx\,dy\,dz} = \frac{-2\,x\,(-2\,z)}{(a^2 - z^2)^2} = \frac{4\,xz}{(a^2 - z^2)^2}.$$

$\dfrac{d^2u}{dz\,dy} = \dfrac{2\,x^2z}{(a^2 - z^2)^2}$.   Consider y, z constant, and differentiate.

$$\frac{d^3u}{dz\,dy\,dx} = \frac{4\,xz}{(a^2 - z^2)^2}.$$

$$\frac{d^2u}{dy\,dx}=\frac{2\,x}{a^2-z^2}.$$ Consider x constant, and differentiate.

$$\frac{d^3u}{dx\,dy\,dz}=\frac{-2\,x\,(-2\,z)}{(a^2-z^2)^2}=\frac{4\,xz}{(a^2-z^2)^2}.$$

Hence $$\frac{d^3u}{dx\,dy\,dz}=\frac{4\,xz}{(a^2-z^2)^2}=\frac{d^3u}{dz\,dy\,dx}=\frac{d^3u}{dy\,dx\,dz}.$$

(6.) $u=x^2y^4$;   find du,   and show that

$$\frac{d^2u}{dy\,dx}=8\,xy^3=\frac{d^2u}{dx\,dy}.$$

(7.) $u=\dfrac{x^5}{y^3}$; $du=\dfrac{x^4}{y^4}\,(5y\,dx-3x\,dy).$

(8.) $u=x^y$; $du=x^y\left(\dfrac{y}{x}dx+\log x\,dy\right),$ and

$$\frac{d^2u}{dy\,dx}=x^y\left(\frac{1}{x}+\frac{y}{x}\log x\right)=\frac{d^2u}{dx\,dy}.$$

(9.) $u=\sin\dfrac{x}{y}$; $\dfrac{d^3u}{dy\,dx^2}=\dfrac{2}{y^3}\sin\dfrac{x}{y}+\dfrac{x}{y^4}\cos\dfrac{x}{y}=\dfrac{d^3u}{dx^2\,dy}.$

(10.) $u=y\sin x+x\sin y$;   show that

$$\frac{d^2u}{dy\,dx}=\cos x+\cos y=\frac{d^2u}{dx\,dy}.$$

(11.) $u=\sin\,(x^2y)$;   show that

$$\frac{d^2u}{dy\,dx}=2\,x\,\{\cos\,(x^2y)-x^2y\,\sin\,(x^2y)\,\}=\frac{d^2u}{dx\,dy}.$$

(12.) $u=\dfrac{xy}{2\,x+z}$;   show that

$$\frac{d^3u}{dz^2\,dy}=\frac{2\,x}{(2\,x+z)^3}=\frac{d^3u}{dy\,dz^2}=\frac{d^3u}{dz\,dy\,dz},$$

$$\frac{d^3u}{dx\,dz^2}=\frac{2\,y\,(z-4\,x)}{(2\,x+z)^4}=\frac{d^3u}{dz^2\,dx}=\frac{d^3u}{dz\,dx\,dz}.$$

(13.) $u = \dfrac{x^2 + y^2}{x + y}$;   find du,   and show that

$$\frac{d^2 u}{dy\,dx} = -4\,\frac{xy}{(x+y)^3} = \frac{d^2 u}{dx\,dy}.$$

(14.) $u = \{(a-x)^2 + (b-y)^2 + (c-z)^2\}^{-\frac{1}{2}}$;   show that

$$\frac{d^2 u}{dx^2} + \frac{d^2 u}{dy^2} + \frac{d^2 u}{dz^2} = 0.$$

(15.) $u = \sin^{-1}\dfrac{x-y}{x}$;   find du,   and show that

$$\frac{d^2 u}{dy\,dx} = \frac{1}{y^{\frac{1}{2}}(2x-y)^{\frac{3}{2}}} = \frac{d^2 u}{dx\,dy}.$$

(16.) $u = \sin^{-1}\dfrac{x^2 - y^2}{x^2 + y^2}$;   show that

$$du = \frac{2}{x^2 + y^2}(y\,dx - x\,dy), \qquad \frac{d^2 u}{dy\,dx} = \frac{2(x^2 - y^2)}{(x^2 + y^2)^2} = \frac{d^2 u}{dx\,dy}.$$

CHAPTER IX.

EULER'S THEOREM FOR THE INTEGRATION OF HOMOGENEOUS FUNCTIONS OF ANY NUMBER OF VARIABLES.

If u be a homogeneous algebraic function of n dimensions of any number of variables x, y, z, &c., then

$$x\,\frac{du}{dx} + y\,\frac{du}{dy} + z\,\frac{du}{dz} + \text{\&c.} = nu.$$

Ex. (1.) $u = \dfrac{x^{\frac{1}{2}} + y^{\frac{1}{2}}}{x + y}$;   here $n = -\dfrac{1}{2}$.

$$\frac{du}{dx} = \frac{(x+y)\cdot\frac{1}{2}x^{-\frac{1}{2}} - (x^{\frac{1}{2}} + y^{\frac{1}{2}})}{(x+y)^2}, \quad \frac{du}{dy} = \frac{(x+y)\cdot\frac{1}{2}y^{-\frac{1}{2}} - (x^{\frac{1}{2}} + y^{\frac{1}{2}})}{(x+y)^2},$$

$$\therefore x\frac{du}{dx}+y\frac{du}{dy}=\frac{\frac{1}{2}(x+y)x^{\frac{1}{2}}-(x^{\frac{1}{2}}+y^{\frac{1}{2}})x+\frac{1}{2}(x+y)y^{\frac{1}{2}}-(x^{\frac{1}{2}}+y^{\frac{1}{2}})y}{(x+y)^2}$$

$$=\frac{\frac{1}{2}(x^{\frac{1}{2}}+y^{\frac{1}{2}})-(x^{\frac{1}{2}}+y^{\frac{1}{2}})}{x+y}=\frac{-\frac{1}{2}(x^{\frac{1}{2}}+y^{\frac{1}{2}})}{x+y}=-\frac{u}{2}.$$

(2.) $u=\sin^{-1}\left(\dfrac{x-y}{x+y}\right)^{\frac{1}{2}}$; here $n=0$. $\sin u=\dfrac{\sqrt{x-y}}{\sqrt{x+y}}$,

$$\cos u\frac{du}{dx}=\frac{\sqrt{x+y}\cdot\dfrac{1}{2\sqrt{x-y}}-\sqrt{x-y}\cdot\dfrac{1}{2\sqrt{x+y}}}{x+y}$$

$$=\frac{y}{(x+y)\sqrt{x^2-y^2}}.$$

$$\cos u=\sqrt{1-\sin^2 u}=\sqrt{1-\frac{x-y}{x+y}}=\frac{\sqrt{2y}}{\sqrt{x+y}},$$

$$\therefore\frac{du}{dx}=\frac{y}{\sqrt{2y}\cdot(x+y)\sqrt{x-y}}.$$

Similarly $\dfrac{du}{dy}=\dfrac{-x}{\sqrt{2y}\,(x+y)\sqrt{x-y}}.$

$$\therefore x\frac{du}{dx}+y\frac{du}{dy}=\frac{xy-xy}{\sqrt{2y}\,(x+y)\sqrt{x-y}}=0.$$

(3.) $u=\sqrt{x^2+y^2}$; here $n=1$.

$$x\frac{du}{dx}+y\frac{du}{dy}=u, \qquad x^2\frac{d^2u}{dx^2}+2xy\frac{d^2u}{dy\,dx}+y^2\frac{d^2u}{dy^2}=0.$$

(4.) $u=(x+y+z)^2$; here $n=2$.

$$x\frac{du}{dx}+y\frac{du}{dy}+z\frac{du}{dz}=2u.$$

(5.) $u=\dfrac{x^2y^2}{x+y}$; here $n=3$. $x\dfrac{du}{dx}+y\dfrac{du}{dy}=3u.$

CHAPTER X.

ELIMINATION OF CONSTANTS AND FUNCTIONS BY DIFFERENTIATION.

Ex. (1.) Let $y - ax^2 + b = 0$; eliminate the constants a and b.

$$\frac{dy}{dx} - 2ax = 0, \qquad a = \frac{dy}{dx} \cdot \frac{1}{2x}.$$

Substituting this value of a in the given equation,

$$y - \frac{dy}{dx} \cdot \frac{x}{2} + b = 0, \qquad \text{an equation from which } a \text{ is eliminated.}$$

To eliminate b, take the equation $\frac{dy}{dx} = 2ax$, and proceed to the second differential coefficient.

$$\frac{d^2y}{dx^2} = 2a. \qquad \text{But } a = \frac{dy}{dx} \cdot \frac{1}{2x},$$

$$\therefore \frac{d^2y}{dx^2} = \frac{dy}{dx} \cdot \frac{1}{x}, \qquad \text{an equation from which } a \text{ and } b \text{ are}$$

both eliminated.

(2.) $y^2 - ax - bx^2 = 0$; eliminate a and b.

$$2y\frac{dy}{dx} = a + 2bx, \qquad \therefore a = 2y\frac{dy}{dx} - 2bx. \quad \dots \quad (1)$$

Differentiating again, we have

$$2y\frac{d^2y}{dx^2} + \frac{dy}{dx} \cdot 2\frac{dy}{dx} = 2b, \qquad \therefore b = y\frac{d^2y}{dx^2} + \left(\frac{dy}{dx}\right)^2. \quad \dots \quad (2)$$

Substituting from (1), (2), the values of a and b in the given equation, there results

$$y^2 = 2xy\frac{dy}{dx} - x^2y\frac{d^2y}{dx^2} - x^2\left(\frac{dy}{dx}\right)^2, \qquad \text{an equation from}$$

which a and b are eliminated.

(3.) If $y = a \sin x + b \sin 2x$; $\quad \dfrac{d^4y}{dx^4} + 5\dfrac{d^2y}{dx^2} + 4y = 0.$

$$\frac{dy}{dx} = a\cos x + 2b\cos 2x, \qquad \frac{d^2y}{dx^2} = -a\sin x - 4b\sin 2x,$$

$$\frac{d^3y}{dx^3} = -a\cos x - 8b\cos 2x,$$

$$\frac{d^4y}{dx^4} = a\sin x + 16b\sin 2x,$$

$$5\frac{d^2y}{dx^2} = -5a\sin x - 20b\sin 2x, \qquad \left.\right\} \quad \therefore \frac{d^4y}{dx^4} + 5\frac{d^2y}{dx^2} + 4y = 0.$$

$$4y = 4a\sin x + 4b\sin 2x.$$

(4.) $y = x^n + ae^{mx}$; $\quad$ eliminate a.

$$\frac{dy}{dx} = nx^{n-1} + ame^{mx}, \qquad \therefore a = \left(\frac{dy}{dx} - nx^{n-1}\right)\cdot\frac{1}{me^{mx}}.$$

Substituting this value of a in the given equation,

$$y = x^n + \left(\frac{dy}{dx} - nx^{n-1}\right)\cdot\frac{1}{m}, \qquad my = mx^n + \frac{dy}{dx} - nx^{n-1},$$

$$\therefore \quad \frac{dy}{dx} - my = nx^{n-1} - mx^n = (n - mx)x^{n-1}.$$

(5.) If $z = xf\left(\dfrac{y}{x}\right) + \phi(xy)$; $\quad x^2\cdot\dfrac{d^2z}{dx^2} - y^2\dfrac{d^2z}{dy^2} = 0.$

First, consider y constant, and differentiate with respect to x.

$$\frac{dz}{dx} = x\left(\frac{-y}{x^2}\right)f'\left(\frac{y}{x}\right) + f\left(\frac{y}{x}\right) + y\phi'(xy)$$

$$= -\frac{y}{x}f'\left(\frac{y}{x}\right) + f\left(\frac{y}{x}\right) + y\phi'(xy),$$

$$\frac{d^2z}{dx^2} = -\frac{y}{x}\left(-\frac{y}{x^2}\right)f''\left(\frac{y}{x}\right) + f'\left(\frac{y}{x}\right)\cdot\left(\frac{y}{x^2}\right) - \frac{y}{x^2}f'\left(\frac{y}{x}\right) + y^2\phi''(xy)$$

$$= \frac{y^2}{x^3}f''\left(\frac{y}{x}\right) + y^2\phi''(xy).$$

Again, $xf\left(\dfrac{y}{x}\right)+\phi(xy)$.　Consider x constant, and differentiate with respect to y.

$$\frac{dz}{dy}=x\cdot\frac{1}{x}f'\left(\frac{y}{x}\right)+x\phi'(xy)=f'\left(\frac{y}{x}\right)+x\phi'(xy),$$

$$\frac{d^2z}{dy^2}=\frac{1}{x}f''\left(\frac{y}{x}\right)+x^2\phi''(xy).$$

$$\therefore\quad x^2\frac{d^2z}{dx^2}=\frac{y^2}{x}f''\left(\frac{y}{x}\right)+x^2y^2\phi''(xy),$$

$$y^2\frac{d^2z}{dy^2}=\frac{y^2}{x}f''\left(\frac{y}{x}\right)+x^2y^2\phi''(xy).$$

Hence　　$x^2\dfrac{d^2z}{dx^2}-y^2\dfrac{d^2z}{dy^2}=0.$

(6.) Let $y=mx^3$; eliminate the constant m, and show that

$$3y=x\frac{dy}{dx}.$$

(7.) Let $y=\sqrt{mx+n}$; eliminate m and n, and show that

$$\left(\frac{dy}{dx}\right)^2=-y\frac{d^2y}{dx^2}.$$

(8.) Let $a+c(cx-y)=0$; eliminate c, and show that

$$y\frac{dy}{dx}=a+x\left(\frac{dy}{dx}\right)^2.$$

(9.) Let $x^2+\dfrac{b}{a^2}y^2=\dfrac{c}{a}$; eliminate the constants a and b,

and show that　　$xy\dfrac{d^2y}{dx^2}+x\left(\dfrac{dy}{dx}\right)^2-y\dfrac{dy}{dx}=0.$

(10.) Let $(a-1)(x+y)-xy+a=0$; eliminate a, and

show that　　$y^2+y+1+(x^2+x+1)\dfrac{dy}{dx}=0.$

(11.) Let $c\tan mx - y\sec mx + a = 0$; eliminate a and c, and show that $\dfrac{d^2y}{dx^2} = -m^2y$.

(12.) Let $y = e^x\cos x$; eliminate the circular and exponential functions, and show that $y = \dfrac{dy}{dx} - \dfrac{1}{2}\dfrac{d^2y}{dx^2}$.

(13.) Let $y = n\cos(rx + \alpha)$; eliminate α and n, and show that $\dfrac{d^2y}{dx^2} = -r^2y$.

(14.) Let $y = \sin(\log x)$; eliminate the functions, and show that $x^2\dfrac{d^2y}{dx^2} + x\dfrac{dy}{dx} + y = 0$.

(15.) Let $y = ae^{2x}\sin(3x + b)$; eliminate a and b, and show that $\dfrac{d^2y}{dx^2} - 4\dfrac{dy}{dx} + 13y = 0$.

(16.) Let $(x-\alpha)^2 + (y-\beta)^2 = r^2$; eliminate α and β, and show that $\dfrac{\left\{1 + \left(\frac{dy}{dx}\right)^2\right\}^3}{\left(\frac{d^2y}{dx^2}\right)^2} = r^2$.

(17.) Let $y = \dfrac{e^x + e^{-x}}{e^x - e^{-x}}$; eliminate the exponentials, and show that $y^2 = 1 - \dfrac{dy}{dx}$.

(18.) Let $\dfrac{z}{x + y^m} = \phi(x^2 - y^2)$; eliminate the arbitrary function ϕ, and show that $y\dfrac{dz}{dx} + x\dfrac{dz}{dy} = mz$.

(19.) Let $\dfrac{1}{n}xz = \phi\dfrac{y}{x}$; eliminate the function ϕ, and show that $x\dfrac{dz}{dx} + y\dfrac{dz}{dy} + z = 0$.

(20.) Let $\dfrac{z-\gamma}{x-\alpha}=\phi\dfrac{y-b}{x-\alpha}$; eliminate the function ϕ, and

show that $\qquad (y-b)\dfrac{dz}{dy}+(z-\gamma)\dfrac{dz}{dx}=x-\alpha.$

CHAPTER XI.

MAXIMA AND MINIMA.

FUNCTIONS OF TWO OR MORE VARIABLES.

If u be a function of two variables x and y, then putting
$\dfrac{du}{dx}=0,\quad \dfrac{du}{dy}=0\ ;\quad$ if $\dfrac{d^2u}{dx^2}\cdot\dfrac{d^2u}{dy^2}>\left(\dfrac{d^2u}{dy\,dx}\right)^2,\quad \dfrac{d^2u}{dx^2}$ and $\dfrac{d^2u}{dy^2}$
having both the same algebraic sign, u will be a maximum
when that sign is *negative*, and a minimum when it is
positive.

If, on substituting the particular values of x and y, determined by putting $\dfrac{du}{dx}=0,\ \dfrac{du}{dy}=0$, in the second differential coefficients, these should vanish, then the third differential coefficients must also vanish, or the function will not
be a maximum or minimum.

If $u=f(x,\ y,\ z)$, then we must put $\dfrac{du}{dx}=0,\ \dfrac{du}{dy}=0$,

$\dfrac{du}{dz}=0$, and we must have the condition fulfilled that

$$\left\{\dfrac{d^2u}{dx^2}\cdot\dfrac{d^2u}{dy^2}-\left(\dfrac{d^2u}{dx\,dy}\right)^2\right\}\cdot\left\{\dfrac{d^2u}{dx^2}\dfrac{d^2u}{dy^2}-\left(\dfrac{d^2u}{dx\,dz}\right)^2\right\}\quad\text{exceeds}$$

$$\left(\dfrac{d^2u}{dy\,dz}\cdot\dfrac{d^2u}{dx^2}-\dfrac{d^2u}{dx\,dy}\cdot\dfrac{d^2u}{dx\,dz}\right)^2.$$

5 / Ex. (1.) Let $u = x^4 + y^4 - 4axy^2$; find x and y when u is a maximum or minimum.

Differentiate, first considering y constant, and then x constant.

$$\frac{du}{dx} = 4x^3 - 4ay^2 = 0, \qquad \frac{du}{dy} = 4y^3 - 8axy = 0,$$

$$\therefore x^3 = ay^2, \qquad y^2 = 2ax, \qquad x^3 = 2a^2x, \qquad x^2 = 2a^2,$$

$$\therefore x = \pm a\sqrt{2}. \qquad y^2 = 2a^2\sqrt{2} = a^2\sqrt{8}, \qquad \therefore y = a\sqrt[4]{8}.$$

$$\frac{d^2u}{dx^2} = 12x^2 = 24a^2,$$

$$\frac{d^2u}{dy^2} = 12y^2 - 8ax = 12a^2\sqrt{8} - 8a^2\sqrt{2} = 16a^2\sqrt{2},$$

$$\frac{d^2u}{dx\,dy} = -8ay = -8a^2\sqrt[4]{8},$$

$$\therefore \frac{d^2u}{dx^2} \cdot \frac{d^2u}{dy^2} > \frac{d^2u}{dx\,dy}, \quad \text{and since the algebraic sign of}$$

$\frac{d^2u}{dx^2}$ and $\frac{d^2u}{dy^2}$ is positive, $x = \pm a\sqrt{2}$, and $y = a \pm \sqrt[4]{8}$, give

$u =$ a minimum.

If we take the values $x = 0$, $y = 0$, then $\frac{d^2u}{dx^2} = 0$, and

$\frac{d^2u}{dy^2} = 0$, and also the third differential coefficients

$$\frac{d^3u}{dx^3} = 24x = 0, \qquad \frac{d^3u}{dy^3} = 24y = 0,$$

Hence also $x = 0$, $y = 0$, give $u =$ a minimum.

3 (2.) To determine the greatest right cone that can be cut out of a given oblate spheroid.

Let $ABDE$ be the ellipse which generates the spheroid, a, b its semi-axes, $CN = x$, $NP = y =$ radius of base of cone.

Then $\quad y^2 = \dfrac{b^2}{a^2}(a^2 - x^2),\quad$ equation to

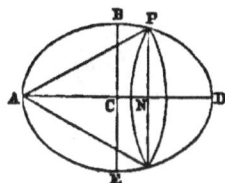

ellipse ;   and

$\qquad \because$ altitude of cone $= AN = a + x,$

$\qquad$ and $\pi y^2 =$ area of base,

$\qquad \therefore$ its volume $v = \tfrac{1}{3}\pi y^2 . (a + x),$ a maximum,

$\qquad \therefore y^2 . (a + x) =$ a maximum,

$$2y\dfrac{dy}{dx}(a+x)+y^2=0, \qquad \therefore \dfrac{dy}{dx}=-\dfrac{1}{2(a+x)}y,$$

$\therefore \dfrac{dy}{dx} = -\dfrac{1}{2(a+x)}\cdot\dfrac{b}{a}\sqrt{a^2-x^2}.\qquad$ But, differentiating the

equation to the ellipse, $y = \dfrac{b}{a}\sqrt{a^2 - x^2},$ we have

$$\dfrac{dy}{dx} = -\dfrac{b}{a}\dfrac{x}{\sqrt{a^2-x^2}}.$$

$$\therefore \dfrac{\sqrt{a^2-x^2}}{2(a+x)} = \dfrac{x}{\sqrt{a^2-x^2}}, \qquad a^2 - x^2 = 2x(a+x),$$

$$a - x = 2x, \qquad \therefore x = \dfrac{a}{3}, \qquad \therefore y^2 = \dfrac{b^2}{a^2}\cdot\dfrac{8a^2}{9} = \dfrac{8}{9}b^2.$$

Hence $\qquad v = \dfrac{1}{3}\pi y^2 . (a+x) = \dfrac{8}{27}\pi b^2 . \dfrac{4a}{3} = \dfrac{32}{81}\pi a b^2.$

4 (3.) Let $u = x^4 + y^4 - 2(x-y)^2$; find the values of x and y which render u a maximum or minimum.

$$\dfrac{du}{dx} = 4x^3 - 4(x-y) = 0, \qquad\qquad \therefore x^3 - (x-y) = 0,$$

$$\dfrac{du}{dy} = 4y^3 + 4(x-y) = 0, \qquad\qquad y^3 + (x-y) = 0,$$

$$x^3 + y^3 = 0, \qquad x^3 - y^3 - 2(x-y) = 0,$$

$$x^3 = -y^3, \qquad \therefore x = -y, \qquad x^3 + x^3 - 2(x+x) = 0,$$

$$2x^3 = 4x, \qquad x^2 = 2, \qquad \therefore x = \pm\sqrt{2}, \qquad y = \mp\sqrt{2}.$$

$$\dfrac{d^2u}{dx^2} = 12x^2 - 4 = 24 - 4 = 20, \qquad \dfrac{d^2u}{dy^2} = 12y^2 - 4 = 24 - 4 = 20,$$

$$\frac{d^2u}{dx\,dy}=4, \qquad \therefore \frac{d^2u}{dx^2}\cdot\frac{d^2u}{dy^2}>\frac{d^2u}{dx\,dy}, \qquad \text{and since the}$$

algebraic sign of $\dfrac{d^2u}{dx^2}$ and $\dfrac{d^2u}{dy^2}$ is positive,

$$\therefore\ x=\pm\sqrt{2},\quad\text{and}\quad y=\mp\sqrt{2},\quad\text{give } u=\text{ a minimum.}$$

(4.) Let $u=a\{\sin x+\sin y+\sin(x+y)\}$; show that u is a maximum when $x=y=60°$.

$$\frac{du}{dx}=a\{\cos x+\cos(x+y)\}=0,\ \frac{du}{dy}=a\{\cos y+\cos(x+y)\}=0,$$

$$\therefore\ x=y,\quad \cos x+\cos(x+y)=\cos x+\cos 2x$$
$$=\cos x+2\cos^2 x-1=0,$$

$$\cos^2 x+\frac{1}{2}\cos x=\frac{1}{2}, \qquad \therefore\cos x=\frac{1}{2}, \qquad x=60°=y.$$

$$\frac{d^2u}{dx^2}=a\{-\sin x-\sin(x+y)\}=-a\{\sin 60+\sin 120\}$$

$$=-a\left\{\frac{\sqrt{3}}{2}+\frac{\sqrt{3}}{2}\right\}=-a\sqrt{3},$$

$$\frac{d^2u}{dy^2}=a\{-\sin y-\sin(x+y)\}=-a\sqrt{3},$$

$$\frac{d^2u}{dx\,dy}=a\{-\sin(x+y)\}=-a\sin 2x=-a\sin 120=-a\,\frac{\sqrt{3}}{2},$$

$$\therefore\frac{d^2u}{dx^2}\cdot\frac{d^2u}{dy^2}>\frac{d^2u}{dx\,dy},\ \text{and}\ \therefore\ \text{the algebraic sign of } \frac{d^2u}{dx^2}\text{ and}$$

$\dfrac{d^2u}{dy^2}$ is negative, $\qquad \therefore u=\dfrac{3a}{2}\sqrt{3}=$ a maximum.

(5.) A cistern, which is to contain a certain quantity of water, is to be constructed in the form of a rectangular parallelopipedon; determine its form, so that the smallest possible expense shall be incurred in lining its internal surface.

Let $a^3 =$ its content, $x =$ length, $y =$ breadth, then $\dfrac{a^3}{xy} =$ depth.

$$\therefore \text{surface} = u = xy + 2\dfrac{a^3}{x} + 2\dfrac{a^3}{y}, \text{ a minimum.}$$

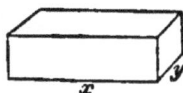

$$\therefore \dfrac{du}{dx} = y - \dfrac{2a^3}{x^2} = 0, \qquad \dfrac{du}{dy} = x - \dfrac{2a^3}{y^2} = 0,$$

$$\therefore x^2 y = xy^2, \qquad x = y, \qquad x^2 y = x^3 = 2a^3, \qquad x = y = 2^{\frac{1}{3}}a.$$

$$\dfrac{a^3}{xy} = \dfrac{a^3}{2^{\frac{2}{3}}a^2} = \dfrac{2^{\frac{1}{3}}a}{2}. \qquad \text{Hence the base must be a square,}$$

and the depth equal to half the length or breadth.

$$\text{Again } \dfrac{d^2u}{dx^2} = \dfrac{4a^3}{x^3} = \dfrac{4a^3}{2a^3} = 2, \qquad \dfrac{d^2y}{dy^2} = 2, \qquad \dfrac{d^2u}{dx\,dy} = 1.$$

$$\therefore \dfrac{d^2u}{dx^2} \cdot \dfrac{d^2u}{dy^2} > \left(\dfrac{d^2u}{dx\,dy}\right)^2. \qquad \text{Hence } u \text{ is a minimum.}$$

(6.) In a given circle to inscribe a triangle whose perimeter shall be the greatest possible.

Let r be the radius, and θ and ϕ two of the angles of the triangle; draw $BD \perp AC$ the base : then, Euc. B. 6. prop. C,

$$c \cdot a = BD \cdot 2r, \qquad \therefore a = 2r \cdot \dfrac{BD}{c} = 2r\sin\theta,$$

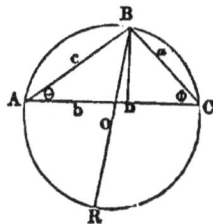

$$\dfrac{c}{a} = \dfrac{\sin\phi}{\sin\theta}, \qquad \therefore c = \dfrac{\sin\phi}{\sin\theta} \times a = 2r\sin\phi,$$

$$\dfrac{b}{a} = \dfrac{\sin B}{\sin\theta} = \dfrac{\sin(\pi - B)}{\sin\theta} = \dfrac{\sin(\theta + \phi)}{\sin\theta},$$

$$\therefore b = \dfrac{\sin(\theta + \phi)}{\sin\theta} \cdot a = 2r\sin(\theta + \phi),$$

Hence $u = a + c + b = 2r\{\sin\theta + \sin\phi + \sin(\theta + \phi)\},$

$$\dfrac{du}{d\theta} = 2r\{\cos\theta + \cos(\theta + \phi)\} = 0,$$

$$\frac{du}{d\phi} = 2r\{\cos\phi + \cos(\theta+\phi)\} = 0,$$

$$\therefore \cos\theta = \cos\phi, \qquad \theta = \phi, \qquad \theta+\phi = 2\theta,$$

$$\therefore \cos\theta + \cos 2\theta = 0, \qquad \cos\theta + 2\cos^2\theta - 1 = 0,$$

$$\cos^2\theta + \frac{1}{2}\cos\theta = \frac{1}{2}, \qquad \therefore \cos\theta = \frac{1}{2}, \qquad \theta = 60° = \phi.$$

Hence the ∠s are all equal, and the △ is equilateral.

$$\frac{d^2u}{d\theta^2} = 2r\{-\sin\theta - \sin(\theta+\phi)\} = -2r\{\sin 60 + \sin 120\}$$

$$= -2r\sqrt{3},$$

$$\frac{d^2u}{d\phi^2} = 2r\{-\sin\phi - \sin(\theta+\phi)\} = -2r\sqrt{3},$$

$$\frac{d^2u}{d\theta\,d\phi} = -2r\{\sin(\theta+\phi)\} = -2r\sin 120 = -2r\cdot\frac{\sqrt{3}}{2} = -r\sqrt{3},$$

$$\therefore \frac{d^2u}{d\theta^2}\cdot\frac{d^2u}{d\phi^2} = 12r^2, \qquad \text{and} \quad \frac{d^2u}{d\theta^2}\cdot\frac{d^2u}{d\phi^2} > \frac{d^2u}{d\theta\,d\phi}.$$

Hence the perimeter is a maximum.

(7.) To determine the least polygon that can be described about a given circle.

Let θ_1, θ_2, θ_3, ... θ_n, be the successive angles contained between the lines from the centre to the angular points of the polygon and the radii of the circle ; then if the radius be r, and the first of those lines be l, the area of the right-angled triangle whose angle at the centre is θ_1 will be

$$\frac{1}{2}rl\sin\theta_1 = \frac{1}{2}r\cdot r\sec\theta_1\cdot\sin\theta_1 = \frac{r^2}{2}\tan\theta_1 ;$$

and similarly of all the n triangles successively, into which the polygon may be supposed to be divided ; so that the entire area of the polygon will be

$$\frac{r^2}{2}(\tan\theta_1 + \tan\theta_2 + \tan\theta_3 + \ldots + \tan\theta_n).$$

But $\tan \theta_n = -\tan\{2\pi - (\theta_1 + \theta_2 + .. + \theta_{n-1})\} = -\tan(2\pi - \phi_1)$,

$\qquad$ where $\phi_1 = \theta_1 + \theta_2 + \ldots + \theta_{n-1}$.

$\therefore\ u = \tan \theta_1 + \tan \theta_2 + \tan \theta_3 + .. - \tan(2\pi - \phi_1)$, a min.

Now, differentiating with respect to θ_1, considering the others constant, and remembering that θ_1 is contained in ϕ_1, the assumed sum of the series, we have

$$\frac{du}{d\theta_1} = \sec^2\theta_1 - \sec^2(2\pi - \phi_1) = 0, \qquad \therefore\ \theta_1 = 2\pi - \phi_1 = \theta_n.$$

And similarly, any one of the angles is equal to the angle immediately preceding; hence all the angles are equal, and the polygon is consequently equilateral.

(8.) Of all triangular pyramids of a given base and altitude, to find that which has the least surface.

Let a, b, c be the sides of the base, h the altitude of the pyramid, θ, ϕ, ψ, the inclination of the faces to the base.

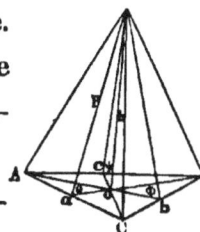

Then, if p be a perpendicular from the vertex on the side a, $\qquad \sin\theta = \dfrac{h}{p}, \qquad \therefore\ p = \dfrac{h}{\sin\theta} = h\ \mathrm{cosec}\ \theta,$

$$\text{area of face} = \frac{1}{2}ap = \frac{1}{2}ah\ \mathrm{cosec}\ \theta,$$

$\therefore$ area of the three faces $= \dfrac{1}{2}ah\ \mathrm{cosec}\,\theta + \dfrac{1}{2}bh\ \mathrm{cosec}\,\phi + \dfrac{1}{2}ch\ \mathrm{cosec}\,\psi,$

$$u = \frac{1}{2}h(a\ \mathrm{cosec}\,\theta + b\ \mathrm{cosec}\,\phi + c\ \mathrm{cosec}\,\psi) \ldots \ldots \ldots (1).$$

Also, the base of the pyramid may be divided into three triangles whose altitudes are readily determined;

$$\because\ \frac{h}{aO} = \tan\theta, \qquad \therefore\ \frac{aO}{h} = \cot\theta, \qquad \therefore\ \text{altitude}\ aO = h\tan\theta,$$

$$\therefore\ \text{area}\ \triangle AOC = \frac{1}{2}a \cdot aO = \frac{1}{2}ah\cot\theta,$$

$$\therefore \text{ area base} = \frac{1}{2}ah\cot\theta + \frac{1}{2}bh\cot\phi + \frac{1}{2}ch\cot\psi,$$

and putting this area $= m^2$, we have

$$m^2 = \frac{1}{2}h\,(a\cot\theta + b\cot\phi + c\cot\psi) \quad \ldots \ldots \text{(2)}.$$

From (1), $\dfrac{du}{d\theta} = \dfrac{h}{2}\left\{ -a\,\operatorname{cosec}\theta\cot\theta - c\,\operatorname{cosec}\psi\cot\psi\,\dfrac{d\psi}{d\theta} \right\} = 0,$

$$\frac{du}{d\phi} = \frac{h}{2}\left\{ -b\,\operatorname{cosec}\phi\cot\phi - c\,\operatorname{cosec}\psi\cot\psi\,\frac{d\psi}{d\phi} \right\} = 0,$$

$$\therefore a\,\operatorname{cosec}\theta\cot\theta = -c\,\operatorname{cosec}\psi\cot\psi\,\frac{d\psi}{d\theta},$$

$$b\,\operatorname{cosec}\phi\cot\phi = -c\,\operatorname{cosec}\psi\cot\psi\,\frac{d\psi}{d\phi},$$

$$a\,\operatorname{cosec}\theta\cot\theta\,\frac{d\psi}{d\phi} = -c\,\operatorname{cosec}\psi\cot\psi\,\frac{d\psi}{d\theta}\cdot\frac{d\psi}{d\phi},$$

$$b\,\operatorname{cosec}\phi\cot\phi\,\frac{d\psi}{d\theta} = -c\,\operatorname{cosec}\psi\cot\psi\,\frac{d\psi}{d\theta}\cdot\frac{d\psi}{d\phi},$$

$$\therefore a\,\operatorname{cosec}\theta\cot\theta\,\frac{d\psi}{d\phi} = b\,\operatorname{cosec}\phi\cot\phi\,\frac{d\psi}{d\theta} \quad \ldots\ldots \text{(3)}.$$

From (2), $\qquad \dfrac{2m^2}{h} = a\cot\theta + b\cot\phi + c\cot\psi,$

$$c\cot\psi = \frac{2m^2}{h} - a\cot\theta - b\cot\phi,$$

$$-c\,(1+\cot^2\psi)\,\frac{d\psi}{d\theta} = a\,(1+\cot^2\theta),$$

$$-c\,(1+\cot^2\psi)\,\frac{d\psi}{d\phi} = b\,(1+\cot^2\phi),$$

$$\therefore \left.\begin{aligned} \frac{d\psi}{d\theta} &= -\frac{a\,\operatorname{cosec}^2\theta}{c\,\operatorname{cosec}^2\psi} \\[2mm] \frac{d\psi}{d\phi} &= -\frac{b\,\operatorname{cosec}^2\phi}{c\,\operatorname{cosec}^2\psi} \end{aligned}\right\} \quad \text{Substitute these values in (3).}$$

$$a \operatorname{cosec} \theta \cot \theta \cdot \frac{b \operatorname{cosec}^2 \phi}{c \operatorname{cosec}^2 \psi} = b \operatorname{cosec} \phi \cot \phi \cdot \frac{a \operatorname{cosec}^2 \theta}{c \operatorname{cosec}^2 \psi},$$

$$\therefore \cot \theta \operatorname{cosec} \phi = \cot \phi \operatorname{cosec} \theta,$$

$$\frac{\cos \theta}{\sin \theta} \cdot \frac{1}{\sin \phi} = \frac{\cos \phi}{\sin \phi} \cdot \frac{1}{\sin \theta}, \qquad \therefore \theta = \phi.$$

Similarly, by finding the partial differential coefficients $\dfrac{du}{d\theta}, \dfrac{du}{d\psi}$, considering first ψ and then θ constant, it may be shown that $\theta = \psi$.

Hence $\theta = \phi = \psi$, or the faces are equally inclined to the base.

ᒋ (9.) Required the dimensions of an open cylindrical vessel of given capacity, so that the smallest possible quantity of metal shall be used in its construction, the thickness of the side and base being already determined upon.

Let a be the given thickness, c the given capacity, $x=$ radius of base inside, $y=$ altitude inside. Then

Whole volume $\mathrm{v} = \pi (x+a)^2 \cdot (y+a)$,

Interior volume $c = \pi x^2 y$,   hence the quantity of metal

$$\mathrm{v} - c = \pi (x+a)^2 \cdot (y+a) - c = \text{a minimum},$$

$$\therefore (x+a)^2 \cdot (y+a) = \text{a minimum}.$$

$$(x+a)^2 dy + (y+a) \cdot 2(x+a) dx = 0, \qquad \therefore \frac{dy}{dx} = -\frac{2(y+a)}{x+a}.$$

But $y = \dfrac{c}{\pi} \cdot \dfrac{1}{x^2}$, $\therefore \dfrac{dy}{dx} = -\dfrac{c}{\pi} \cdot \dfrac{2}{x^3}$, $\therefore \dfrac{\frac{c}{\pi x^2} + a}{x+a} = \dfrac{c}{\pi x^3}$,

Whence $x = y = \left(\dfrac{c}{\pi}\right)^{\frac{1}{3}}$. Therefore the altitude must be made equal to the radius of the base.

(10.) $u = x^3 - 3axy + y^3$; find the values of x and y which render u a maximum or minimum.

$x = a$, $y = a$, $u = a$ minimum when a is positive,

and a maximum when a is negative.

(11.) $u = ax^3 - bx^2y + y^2$; find the values of x and y which make u a maximum or minimum.

(12.) $u = ax^3y^2 - x^4y^2 - x^3y^3$; find the values of x and y which make u a maximum or minimum.

$$x = \frac{a}{2}, \quad y = \frac{a}{3}, \quad u = \frac{a^6}{432} \text{ a maximum.}$$

(13.) $u = \left(1 - \dfrac{a}{x} - \dfrac{b}{y}\right) \cdot \left(1 - \dfrac{x+y}{c}\right)$; find the values of x and y which render u a maximum or minimum.

(14.) $u = a \cos^2 x + b \cos^2 y$, where $y = \dfrac{\pi}{4} + x$; find the values of $\cos x$ and $\cos y$ which make u a maximum or minimum.

$$\cos^2 x = \frac{1}{2} \pm \frac{a}{2\sqrt{a^2 + b^2}}, \qquad \cos^2 y = \frac{1}{2} \pm \frac{b}{2\sqrt{a^2 + b^2}},$$

$u = \dfrac{1}{2}(a + b \pm \sqrt{a^2 + b^2})$, a maximum with the upper, and a minimum with the lower sign.

(15.) Divide a given number a into three such parts x, y, and z, that $\dfrac{xy}{2} + \dfrac{xz}{3} + \dfrac{yz}{4}$ shall be a maximum or minimum, and determine which it is.

(16.) Inscribe the greatest triangle within a given circle.

The triangle is equilateral.

(17.) A given sphere is to be formed into a solid composed of two equal cones on opposite sides of a common base, in such a manner that its surface may be the least possible: find the dimensions of the solid, and compare its surface with that of the sphere.

(18.) Show that the greatest polygon that can be inscribed in a given circle is a regular polygon.

(19.) In a given ellipsoid, whose equation is $\frac{x^2}{a^2}+\frac{y^2}{b^2}+\frac{z^2}{c^2}=1$, to inscribe the greatest parallelopipedon.

If x, y, z be the half-edges of the parallelopipedon,

$$x=\frac{a}{\sqrt{3}},\ y=\frac{b}{\sqrt{3}},\ z=\frac{c}{\sqrt{3}},\ u=\frac{8abc}{3^{\frac{3}{2}}}.$$

(20.) To find a point P within a given triangle, from which, if lines be drawn to the angular points, the sum of their squares shall be a minimum.

If A, B, C be the angles, a, b, c the sides of the triangle ; then $CP=\frac{1}{3}(2a^2+2b^2-c^2)^{\frac{1}{2}}$.

The point is the centre of gravity of the triangle.

(21.) Divide the quadrant of a circle into three parts, such that the sum of the products of the sines of every two shall be a maximum or minimum, and determine which it is.

CHAPTER XII.

TANGENTS, NORMALS, AND ASYMPTOTES TO CURVES.

If $y=f(x)$ be the equation to a curve,

$y'-y=\frac{dy}{dx}(x'-x)$ is the equation to a tangent.

If $u=\phi(x, y)=c$ be the equation to the curve,

$\frac{du}{dx}(x'-x)+\frac{du}{dy}(y'-y)=0$ is the equation to the tangent.

The equations to the normal are

$$y'-y=-\frac{dx}{dy}(x'-x), \quad \text{and} \quad \frac{du}{dx}(y'-y)-\frac{du}{dy}(x'-x)=0.$$

$$\text{The tangent}=y\sqrt{1+\left(\frac{dx}{dy}\right)^2}, \quad \text{Normal}=y\sqrt{1+\left(\frac{dy}{dx}\right)^2},$$

$$\text{Subtangent}=y\frac{dx}{dy}, \quad \text{Subnormal}=y\frac{dy}{dx}.$$

The portion of the axis of y intercepted between the origin and the tangent is $y-x\frac{dy}{dx}=y_0$.

The portion of the axis of x so intercepted is $x-y\frac{dx}{dy}=x_0$.

Ex. (1.) Draw a tangent and normal to a given point P in the common or conical parabola.

$y^2=4ax$ is the equation to the curve,

$$2y\frac{dy}{dx}=4a, \quad \therefore \frac{dy}{dx}=\frac{2a}{y}.$$

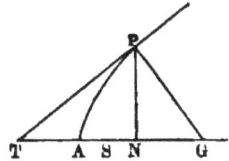

$$\text{Subtangent } NT=y\frac{dx}{dy}=\frac{y^2}{2a}=2x.$$

Hence to draw the tangent, let fall the perpendicular PN, take $NT=2AN$, and join PT; PT will be the tangent.

$$\text{Subnormal } NG=y\frac{dy}{dx}=2a.$$

Hence to draw the normal, take $NG=2AS$, and join PG; PG will be the normal.

(2.) Let $y^n=a^{n-1}x$ be the equation to a curve; find the subnormal and subtangent.

$$ny^{n-1}\cdot\frac{dy}{dx}=a^{n-1}, \quad \therefore \frac{dy}{dx}=\frac{a^{n-1}}{ny^{n-1}},$$

$$\text{Subnormal } NG=y\frac{dy}{dx}=\frac{ya^{n-1}}{ny^{n-1}}=\frac{a^{n-1}}{ny^{n-2}}=\frac{\frac{y^n}{x}}{ny^{n-2}}=\frac{y^2}{nx};$$

Subtangent $NT = y\dfrac{dx}{dy} = \dfrac{ny^n}{a^{n-1}} = \dfrac{ny^n}{\dfrac{y^n}{x}} = nx.$

If $n=2$, $y^2 = ax$, $NG = \dfrac{a}{2}$, $NT = 2x$, and the curve is a parabola.

(3.) Let $u = x^3 - 3axy + y^3 = 0$ be the equation to a curve; determine the subtangent.

$$\dfrac{du}{dx} = 3x^2 - 3ay - 3ax\dfrac{dy}{dx} + 3y^2\dfrac{dy}{dx} = 0,$$

$$\therefore (y^2 - ax)\dfrac{dy}{dx} = ay - x^2, \qquad \dfrac{dy}{dx} = \dfrac{ay - x^2}{y^2 - ax},$$

$$\therefore \text{Subtangent } NT = y\dfrac{dx}{dy} = \dfrac{y^3 - axy}{ay - x^2}.$$

(4.) If $y^2 = 4a(x+a)$ be the equation to a parabola, the origin in the focus; show that the points of intersection of the tangents with perpendiculars from the focus are determined by the equations $x_1 = -a$, $y_1 = \dfrac{y}{2}.$

S the focus, $AS = a$, $SN = x$, $AN = x+a$, $NP = y$,

$y^2 = 4a(x+a) \;\ldots\; (1),$ eq^n. to curve,

$y_1 - y = \dfrac{dy}{dx}(x_1 - x) \;..\; (2),$ eq^n. to tan.,

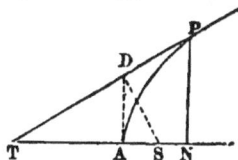

$y_1 = -\dfrac{dx}{dy}x_1, \;\ldots\ldots\; (3),$ eq^n. to ppdr. from origin,

$\therefore$ by subtraction, $y = -\dfrac{dx}{dy}x_1 - \dfrac{dy}{dx}x_1 + \dfrac{dy}{dx}x, \;\ldots\ldots\; (4).$

(1) $\dfrac{dy}{dx} = \dfrac{2a}{y}, \qquad \dfrac{dx}{dy} = \dfrac{y}{2a}, \qquad x+a = \dfrac{y^2}{4a}, \qquad x = \dfrac{y^2}{4a} - a,$

(4) $\left(\dfrac{dy}{dx} + \dfrac{dx}{dy}\right)x_1 = \dfrac{dy}{dx}x - y, \qquad \therefore$ by substitution

$$\left(\frac{2a}{y}+\frac{y}{2a}\right)x_{,}=\frac{2a}{y}\left(\frac{y^2}{4a}-a\right)-y,$$

$$\frac{4a^2+y^2}{2ay}x_{,}=\frac{y}{2}-\frac{2a^2}{y}-y=-\frac{2a^2}{y}-\frac{y}{2}=-\frac{4a^2+y^2}{2y},$$

$$\therefore \frac{1}{a}x_{,}=-1, \qquad x_{,}=-a,$$

$$y_{,}=-\frac{dx}{dy}x_{,}=-\frac{y}{2a}\cdot(-a)=\frac{y}{2}.$$

(5.) The equation $x^m y^n = a$, which includes the common hyperbola, is said to belong to hyperbolas of all orders. Find the subtangent at a given point in the curve.

$$x^m = \frac{a}{y^n},$$

$$mx^{m-1}\cdot\frac{dx}{dy}=-\frac{any^{n-1}}{y^{2n}}=-\frac{an}{y^{n+1}}, \qquad \frac{dx}{dy}=-\frac{an}{mx^{m-1}\cdot y^{n+1}},$$

$$\therefore \text{Subtan. } NT=y\frac{dx}{dy}=-\frac{n}{mx^{m-1}}\cdot\frac{a}{y^n}=-\frac{n}{mx^{m-1}}\cdot x^m=-\frac{n}{m}x.$$

(6.) Given two points A and B, find the locus of P when the angle PBA is double of the angle PAB, and draw an asymptote to the curve traced by P.

A the origin, $AB=a$, $AN=x$, $NP=y$, $A=\theta$, $B=2\theta$.

$$\frac{PN}{AN}=\frac{y}{x}=\tan\theta, \qquad \frac{PN}{BN}=\frac{y}{a-x}=\tan B=\tan 2A=\frac{2\tan\theta}{1-\tan^2\theta}.$$

$$\therefore \frac{y}{a-x}=\frac{2\cdot\frac{y}{x}}{1-\frac{y^2}{x^2}}=\frac{2xy}{x^2-y^2}$$

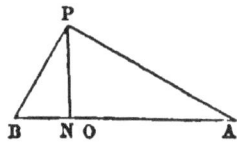

$\therefore y^2=3x^2-2ax,$ the equation to the curve.

Whence, if $y=0$, $x=\frac{2}{3}a$, and taking $AO=\frac{2}{3}AB$, the curve will pass through O.

The origin may be changed to O by putting $x_1 = ON$, and substituting the resulting value of x in the equation to the curve ; whence

$$y = \pm x \left(3 + \frac{2a}{x}\right)^{\frac{1}{2}},$$

$$y = \pm x \left\{3^{\frac{1}{2}} + \frac{1}{2} \cdot 3^{-\frac{1}{2}} \cdot \frac{2a}{x} + \frac{\frac{1}{2}\left(\frac{1}{2} - 1\right)}{2} \cdot 3^{-\frac{3}{2}} \cdot \left(\frac{2a}{x}\right)^2 - \frac{D}{x^3}\&c.\right\}$$

$$= \pm 3^{\frac{1}{2}}x \pm \frac{a}{3^{\frac{1}{2}}} \mp \frac{a^2}{2 \cdot 3^{\frac{3}{2}}x} \mp \frac{D}{x^2} \mp \&c.$$

$\therefore y = \pm x \sqrt{3} \pm \dfrac{a}{\sqrt{3}}$ is the equation to the asymptote.

If $x = 0$, $\quad y = \pm \dfrac{a}{\sqrt{3}}$; $\quad$ if $y = 0$, $\quad x = \mp \dfrac{a}{3}$,

$\therefore \dfrac{dy}{dx} = \sqrt{3} = \tan 60°$, and the asymptote cuts the axis of

x at an $\angle$ of $60°$, and at a distance $= -\dfrac{a}{3}$ from the point O.

(7.) If $y^2 = \dfrac{x^3 + ax^2}{x - a}$ be the equation to a curve ; find the equation to the asymptote.

$$y^2 = x^2 \left(\frac{x+a}{x-a}\right) = x^2 \left(1 + \frac{2a}{x} + \frac{2a^2}{x^2} + \&c.\right)$$

$$\therefore y = \pm x \left(1 + \frac{a}{x} + \frac{a^2}{x^2} + \&c.\right)$$

$\therefore y = \pm (x + a)$ is the equation to two asymptotes, and $\because$ if $x = 0$, $y = a$, $\therefore$ an asymptote cuts the axis of y at the distance a from the origin ; and $\because$ if $y = 0$, $x = -a$, $\therefore$ an asymptote cuts the axis of x at the distance $-a$ from the origin.

Again $\because \dfrac{dy}{dx} = \pm 1 = \tan 45°$ or $\tan 135°$, $\therefore$ these asymptotes cut the axes at an angle of 45°, and are consequently at right-angles to each other.

Putting $x=a$ in the equation to the curve, we have

$$y^2 = \frac{2x^3}{0}, \text{ or } y = \frac{x^{\frac{3}{2}}\sqrt{2}}{0} = \infty,$$

∴ there is another asymptote parallel to the axis of y.

(8.) If $y-2=(x-1)\sqrt{x-2}$ be the equation to a curve ; find the point and angle at which the curve cuts the axis of x, and the values of x and y when the tangent is perpendicular to that axis.

If $x=0$, $y-2=-\sqrt{-2}$, ∴ $y=2-\sqrt{-2}$.

If $y=0$, $(x-1)\sqrt{x-2}=-2$, $(x^2-2x+1)(x-2)=4$,

$$x^3-4x^2+5x-6=0,$$

$$x^3-3x^2-x^2+3x+2x-6=0,$$

$$x^2(x-3)-x(x-3)+2(x-3)=0, \qquad \therefore x=3.$$

$$\frac{dy}{dx}=(x-1)\cdot\frac{1}{2\sqrt{x-2}}+\sqrt{x-2}=\frac{x-1+2x-4}{2\sqrt{x-2}}=\frac{3x-5}{2\sqrt{x-2}}.$$

Hence, if $x=3$, $\frac{dy}{dx}=\tan\theta=\frac{9-5}{2\sqrt{3-2}}=\frac{4}{2}=2$, and the

curve cuts the axis of x at a distance 3 from the origin, and at an angle whose tangent is 2.

Again, if $x=2$, $\sqrt{x-2}=0$, ∴ $y-2=0$, $y=2$,

$$\frac{dy}{dx}=\frac{3x-5}{2\sqrt{x-2}}=\frac{6-5}{0}=\frac{1}{0}=\infty \text{ when } x=2.$$

Hence the tangent cuts the axis of x at an angle of 90°, or it is perpendicular to that axis when $x=2$ and $y=2$.

(9.) If from any point P in an ellipse a straight line be drawn to the centre making an angle θ with the normal, and if l be the inclination of the normal to the axis major ; show that $\tan\theta=\dfrac{\tan l\,(a^2-b^2)}{a^2+b^2\tan^2 l}.$

Let $CA=a$, $CB=b$, $CN=x$, $NP=y$,

$\quad\angle CPG=\theta$, $\quad CGP=l$.

$y^2=\dfrac{b^2}{a^2}(a^2-x^2)=b^2-\dfrac{b^2x^2}{a^2}$, eqⁿ. to ellipse.

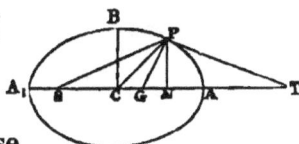

$$NG=\frac{b^2}{a^2}CN=\frac{b^2x}{a^2}, \quad\text{by a property of the ellipse,}$$

$$\tan l=\frac{y}{NG}=\frac{y}{\dfrac{b^2x}{a^2}}=\frac{a^2y}{b^2x},$$

$$\therefore \frac{y}{x}=\frac{b^2}{a^2}\tan l, \quad\text{also }\frac{y}{x}=\tan PCN=\frac{b^2}{a^2}\tan l,$$

$$\theta=CPG=PGN-PCN,$$

$$\tan\theta=\tan(PGN-PCN)=\frac{\tan PGN-\tan PCN}{1+\tan PGN\cdot\tan PCN}$$

$$=\frac{\tan CGP-\tan PCN}{1+\tan CGP\cdot\tan PCN}=\frac{\tan l-\dfrac{b^2}{a^2}\tan l}{1+\tan l\cdot\dfrac{b^2}{a^2}\tan l}$$

$$=\frac{a^2\tan l-b^2\tan l}{a^2+b^2\cdot\tan^2 l}=\frac{\tan l\,(a^2-b^2)}{a^2+b^2\cdot\tan^2 l}.$$

(10.) From the centre C of a circle a radius CR is drawn cutting the chord BD in M, MP is drawn at right-angles to BD and equal to MR; determine the locus of P, and draw the asymptotes.

Let BD, CO be the co-ordinate axes, A the origin,

$CR=a$, $CA=c$, $AM=x$, $MP=y$. Then

$$MP=MR=CR-CM$$

$$=CR-\sqrt{CA^2+AM^2}, \text{ or}$$

$$y=a-\sqrt{c^2+x^2}, \text{ the equation required.}$$

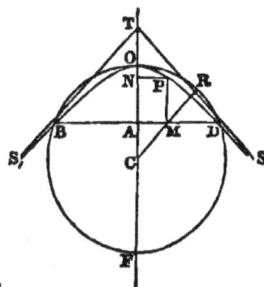

If $x=0$, $y=a-c=CR-CA=CO-CA=AO$.

If $y=0$, $x=\sqrt{a^2-c^2}=\sqrt{CF^2-CA^2}$

$$=\sqrt{(CF+CA)(CF-CA)}=\sqrt{AF\cdot AO}=\sqrt{AD^2},$$

$$\therefore x=AD \text{ or } AB.$$

If $x=\pm\infty$, $y=-\infty$. Hence the curve passes from O through B and D to infinity.

To determine the direction of the tangents at these three points; $\dfrac{dy}{dx}=\tan\theta=\mp\dfrac{x}{\sqrt{c^2+x^2}}=0$ if $x=0$, $\therefore$ at O the tangent is parallel to the axis of x.

$\dfrac{dy}{dx}=\tan\theta=\mp\dfrac{x}{\sqrt{c^2+x^2}}=\dfrac{AD}{\sqrt{CA^2+AD^2}}=\dfrac{\sqrt{a^2-c^2}}{a}$, which determines the direction of the tangents at D and B.

Again, putting $x_1=ON=CO-CN=a-(c+y)$, we have $y=a-c-x_1$; and putting $y_1=NP=x$; and substituting these values of x and y in the equation to the curve, the origin will be transferred to O. Thus

$$a-c-x_1=a-\sqrt{c^2+y_1^2}, \qquad \sqrt{c^2+y_1^2}=c+x_1,$$

$$c^2+y_1^2=c^2+2cx_1+x_1^2,$$

$\therefore y_1^2=2cx_1+x_1^2$, which is the equation to the rectangular hyperbola.

To find the equation to its asymptotes,

$y=(x^2+2cx)^{\frac{1}{2}}$

$$=\pm\left\{(x^2)^{\frac{1}{2}}+\frac{1}{2}(x^2)^{-\frac{1}{2}}(2cx)+\frac{\frac{1}{2}\left(\frac{1}{2}-1\right)}{2}(x^2)^{-\frac{3}{2}}(2cx)^2+\&\text{c.}\right\}$$

$$=\pm\left\{x+c-\frac{c^2}{2x}-\frac{D}{x^2}-\&\text{c.}\right\}.$$

$\therefore y=\pm(x+c)$ is the equation to the two asymptotes;
and $\because$ putting $y=0$, we have $x=-c$, and
putting $x=0$, we have $y=\pm c$; also

$\because \dfrac{dy}{dx}=\tan\theta=\pm1$; $\therefore$ the asymptotes cut the axis OF

at $\angle s=45°$ and $315°$, at the distance $-c$ from the origin O.

Take $OT=CA$, and draw the lines TS, TS, at $\angle s=45°$
and $315°$ respectively, these will be the asymptotes.

(11.) The normal to the curve whose equation is $y^2=4ax$,

is a tangent to the curve defined by $y^2=\dfrac{4}{27a}(x-2a)^3$.

$y^2=4ax,\quad \dfrac{dy}{dx}=\dfrac{2a}{y},\quad \therefore y_,-y=-\dfrac{y}{2a}(x_,-x)$, eq^n. to normal,

$\therefore y_,=-\dfrac{y}{2a}x_,+y\left(\dfrac{x+2a}{2a}\right)$.   Let $y_,=0$, then

$x_,=x+2a=$part cut off from axis of x.

$y^2=\dfrac{4}{27a}(x-2a)^3,\quad 2\log y=\log\dfrac{4}{27a}+3\log(x-2a),$

$2\dfrac{dy}{dx}.\dfrac{1}{y}=3.\dfrac{1}{x-2a},\quad y\dfrac{dx}{dy}=\dfrac{2}{3}(x-2a),$

$\therefore x-y\dfrac{dx}{dy}=x-\dfrac{2}{3}(x-2a)=\dfrac{x+4a}{3}=$ part cut off from
axis of x.

Hence, that the normal and tangent may cut the axis of x at the same point, we must have the equation

$\dfrac{x+4a}{3}=x+2a,\qquad \therefore x=3x+2a.$

But, the angles they make with the axis of x ought to be the same, and since

$\dfrac{dx}{dy}=\dfrac{y}{2a},\qquad$ and $\dfrac{dy}{dx}=\dfrac{3}{2}.\dfrac{y}{x-2a},$

$$\therefore \frac{y}{2a}=\frac{3}{2}\cdot\frac{y}{x-2a}, \qquad \frac{y^2}{a^2}=\frac{4x}{a}=\frac{9\,y^2}{(x-2a)^2}=\frac{4}{3a}(x-2a),$$

$$\therefore 3x=x-2a, \qquad x=3x+2a, \qquad \text{the same as before.}$$

Hence, the normal and tangent, cutting the axis of x in the same point and at the same angle, must be coincident.

(12.) In the curve defined by $y^3=ax^2+x^3$ prove that the portion of the axis of y intercepted between the origin and the tangent $=\dfrac{a}{3}\cdot\left(\dfrac{x}{a+x}\right)^{\frac{2}{3}}$.

$$3y^2\frac{dy}{dx}=2ax+3x^2, \qquad \frac{dy}{dx}=\frac{2ax+3x^2}{3y^2},$$

$$AD=y_0=y-x\frac{dy}{dx}=y-\frac{2ax^2+3x^3}{3y^2}=\frac{3y^3-2ax^2-3x^3}{3y^2}$$

$$=\frac{3(y^3-x^3)-2ax^2}{3y^2}=\frac{ax^2}{3y^2}=\frac{a}{3}\cdot\frac{x^2}{(ax^2+x^3)^{\frac{2}{3}}}$$

$$=\frac{a}{3}\cdot\frac{x^2}{\{(a+x)\,x^2\}^{\frac{2}{3}}}=\frac{a}{3}\cdot\frac{x^2}{(a+x)^{\frac{2}{3}}x^{\frac{4}{3}}}=\frac{a}{3}\cdot\left(\frac{x}{a+x}\right)^{\frac{2}{3}}.$$

(13.) If $y^{\frac{2}{3}}=a^{\frac{2}{3}}-x^{\frac{2}{3}}$; draw a tangent to the curve, and show that the part of the tangent intercepted between the axes $=a$, and that perpendicular on tangent $=\sqrt[3]{axy}$.

$$y^{\frac{2}{3}}=a^{\frac{2}{3}}-x^{\frac{2}{3}}, \qquad x^{\frac{2}{3}}=a^{\frac{2}{3}}-y^{\frac{2}{3}},$$

$$\frac{2}{3}\cdot y^{-\frac{1}{3}}\cdot\frac{dy}{dx}=-\frac{2}{3}x^{-\frac{1}{3}}, \qquad \frac{dy}{dx}=-\frac{y^{\frac{1}{3}}}{x^{\frac{1}{3}}},$$

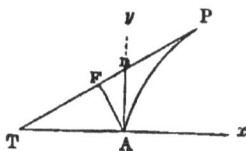

$$x\frac{dy}{dx}=-x^{\frac{2}{3}}y^{\frac{1}{3}}=-y^{\frac{1}{3}}(a^{\frac{2}{3}}-y^{\frac{2}{3}})=y-a^{\frac{2}{3}}y^{\frac{1}{3}}.$$

$$\therefore AD=y-x\frac{dy}{dx}=a^{\frac{2}{3}}y^{\frac{1}{3}}.$$

$$AT=y\frac{dx}{dy}-x=y\left(-\frac{x^{\frac{1}{3}}}{y^{\frac{1}{3}}}\right)-x=-x^{\frac{1}{3}}y^{\frac{2}{3}}-x$$

$$=-x^{\frac{1}{3}}(y^{\frac{2}{3}}+x^{\frac{2}{3}})=-x^{\frac{1}{3}}a^{\frac{2}{3}}.$$

Now $DT^2 = AD^2 + AT^2 = a^{\frac{4}{3}} y^{\frac{2}{3}} + a^{\frac{4}{3}} x^{\frac{2}{3}} = a^{\frac{4}{3}} (y^{\frac{2}{3}} + x^{\frac{2}{3}})$

$$= a^{\frac{4}{3}} \cdot a^{\frac{2}{3}} = a^2,$$

$\therefore DT = a =$ part of tan. intercepted between the axes.

Again $\dfrac{FD}{AD} = \dfrac{AD}{DT}$, $\qquad FD = \dfrac{AD^2}{DT} = \dfrac{a^{\frac{4}{3}} y^{\frac{2}{3}}}{a} = a^{\frac{1}{3}} y^{\frac{2}{3}}.$

$AF^2 = AD^2 - FD^2 = a^{\frac{4}{3}} y^{\frac{2}{3}} - a^{\frac{2}{3}} y^{\frac{4}{3}} = a^{\frac{2}{3}} y^{\frac{2}{3}} (a^{\frac{2}{3}} - y^{\frac{2}{3}}) = a^{\frac{2}{3}} y^{\frac{2}{3}} x^{\frac{2}{3}},$

$\therefore AF = a^{\frac{1}{3}} x^{\frac{1}{3}} y^{\frac{1}{3}} =$ length of perpendicular on tangent.

(14.) Suppose a rigid rod BP slides along the line Ax in such a manner that its extremity P shall be constantly in a given curve whose equation is $y = f(x)$, and let BQ be an nth part of BP; determine the equation to the locus of Q.

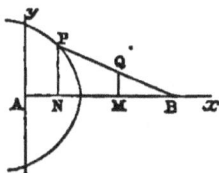

Let $BP = a$, $AN = x$, $NP = y$, $AM = x_{,}$, $MQ = y_{,}$. Then

$$MQ : NP :: BQ : BP, \quad \text{or} \quad y_{,} : y :: \frac{a}{n} : a,$$

$$\therefore y_{,} = \frac{1}{n} y = \frac{1}{n} \cdot f(x).$$

But $AN = AM - NM = AM - (NB - MB) = x_{,} - (nMB - MB)$

$$= x_{,} - (n-1) MB = x_{,} - (n-1) \sqrt{\frac{a^2}{n^2} - y_{,}^2},$$

$\therefore y_{,} = \dfrac{1}{n} \cdot f \left\{ x_{,} - \dfrac{n-1}{n} \sqrt{a^2 - n^2 y_{,}^2} \right\}$, the equation required.

(15.) Determine the subtangent to the curve of which the normal $= 2 a^2 \cdot (\text{abscissa})^3$.

Let x be its abscissa, y its ordinate. Then

$\because$ Normal $PG = y \dfrac{dy}{dx}$, $\qquad \therefore y \dfrac{dy}{dx} = 2 a^2 x^3$, an equation

evidently derivable by differentiation from $\dfrac{y^2}{2} = \dfrac{a^2 x^4}{2}$,

$\therefore y = a x^2$ is the equation to the curve.

Now $\because \dfrac{dy}{dx} = \dfrac{2a^2x^3}{y} = \dfrac{2a^2x^3}{ax^2} = 2ax,$ $\qquad \therefore \dfrac{dx}{dy} = \dfrac{1}{2ax},$

$$\therefore \text{Subtangent } NT = y\dfrac{dx}{dy} = \dfrac{ax^2}{2ax} = \dfrac{x}{2}.$$

The equation to the curve may be put into the form $x^2 = \dfrac{1}{a}\,y$, therefore the curve is a parabola, whose parameter is $\dfrac{1}{a}$, and whose line of abscissæ is perpendicular to the horizontal axis.

(16.) The equation to the catenary is $2y = c\left(e^{\frac{x}{c}} + e^{-\frac{x}{c}}\right)$; find the length of the normal.

$$\dfrac{dy}{dx} = \dfrac{c}{2}\left\{e^{\frac{x}{c}}\left(\dfrac{1}{c}\right) + e^{-\frac{x}{c}}\left(-\dfrac{1}{c}\right)\right\} = \dfrac{1}{2}\{e^{\frac{x}{c}} - e^{-\frac{x}{c}}\},$$

$$\dfrac{dy^2}{dx^2} = \dfrac{e^{\frac{2x}{c}} - 2 + e^{-\frac{2x}{c}}}{4}, \qquad 1 + \dfrac{dy^2}{dx^2} = 1 + \dfrac{e^{\frac{2x}{c}} - 2 + e^{-\frac{2x}{c}}}{4},$$

$$\therefore \sqrt{1 + \dfrac{dy^2}{dx^2}} = \sqrt{\dfrac{e^{\frac{2x}{c}} + 2 + e^{-\frac{2x}{c}}}{2}} = \dfrac{e^{\frac{x}{c}} + e^{-\frac{x}{c}}}{2} = \dfrac{y}{c}.$$

$$\therefore \text{normal } PG = y\sqrt{1 + \dfrac{dy^2}{dx^2}} = y \cdot \dfrac{y}{c} = \dfrac{1}{c}y^2.$$

(17.) If $y^n - (a+bx)\,y^{n-1} + (c + ex + fx^2)\,y^{n-2} - \&c. = 0$ be the equation to a curve of n dimensions, prove that, if each ordinate be divided by the corresponding subtangent, the sum of the quotients will be a constant quantity.

Let $r_1,\ r_2,\ r_3,\ \ldots r_n$ be the values of y which satisfy the given equation, and

$s_1,\ s_2,\ s_3,\ \ldots s_n$ the subtangents corresponding to these values of y; then, by the theory of equations,

$$r_1 + r_2 + r_3 \ldots r_n = a + bx,$$

$$\therefore \frac{dr_1}{dx}+\frac{dr_2}{dx}+\frac{dr_3}{dx}\ldots+\frac{dr_n}{dx}=b;$$

and, taking the differential expression for the subtangents,

$$s_1=\frac{r_1 dx}{dr_1}, \qquad s_2=\frac{r_2 dx}{ar_2}, \quad\ldots\quad s_n=\frac{r_n dx}{dr_n}.$$

$$\therefore \frac{r_1}{s_1}=\frac{dr_1}{dx}, \qquad \frac{r_2}{s_2}=\frac{dr_2}{dx}, \quad\ldots\quad \frac{r_n}{s_n}=\frac{dr_n}{dx}.$$

Hence $\dfrac{r_1}{s_1}+\dfrac{r_2}{s_2}+\dfrac{r_3}{s_3}\ldots+\dfrac{r_n}{s_n}=b.$

(18.) If $y^4-x^4+2bx^2y=0$ be the equation to a curve; find the equation to the asymptote.

Assume $y=xz$, then $x^4z^4-x^4+2bx^3z=0$,

$$x=\frac{2bz}{1-z^4}, \qquad y=\frac{2bz^2}{1-z^4}, \qquad \text{which both become infinite}$$

when $z^4=1$ or $z=1$.

$$4y^3\frac{dy}{dx}-4x^3+2bx^2\frac{dy}{dx}+2by\cdot 2x=0,$$

$$(2y^3+bx^2)\frac{dy}{dx}=2x^3-2bxy, \qquad \frac{dy}{dx}=\frac{2x^3-2bxy}{2y^3+bx^2},$$

$$AD=y-x\frac{dy}{dx}=y-\frac{2x^4-2bx^2y}{2y^3+bx^2}=\frac{2y^4+bx^2y-2x^4+2bx^2y}{2y^3+bx^2}$$

$$=\frac{2(y^4-x^4)+3bx^2y}{2y^3+bx^2}=\frac{-4bx^2y+3bx^2y}{2y^3+bx^2}=-\frac{bx^2y}{2y^3+bx^2},$$

$$=-\frac{bxz}{2xz^3+b}, \qquad \text{which, when } z=1, \text{ and consequently}$$

$x=\infty$, becomes $AD=-\dfrac{bx}{2x+b}=-\dfrac{b}{2+\dfrac{b}{x}}=-\dfrac{b}{2}.$

Hence $y=x-\dfrac{b}{2}$, $y=-x-\dfrac{b}{2}$ are the equations to two asymptotes.

(19.) Investigate an expression for the subtangent : and in the parabola of the nth order, whose equation is $y=ax^n$, find the subtangent and subnormal.

$$\text{Subtangent}=\frac{1}{n}\,x, \quad \text{subnormal}=na^2x^{2n-1}.$$

(20.) The equation to the ellipse being $y^2=\dfrac{b^2}{a^2}(2\,ax-x^2)$; find the subtangent and subnormal.

$$\text{Subtangent}=\frac{2\,ax-x^2}{a-x}, \quad \text{subnormal}=\frac{b^2}{a^2}(a-x).$$

(21.) Prove that $\dfrac{dy}{dx}$ equals the tangent of the angle at which a curve, referred to rectangular co-ordinates, is inclined to the axis.

(22.) $y^2=a^2-x^2$ being the equation to the circle, the origin at the centre, show that the curve cuts the axis of x at an angle of 90°.

(23.) $y^2=2\,ax-x^2$ being the equation to the circle, the origin in the circumference, find the subtangent and normal.

$$\text{Subtangent}=\frac{2\,ax-x^2}{a-x}, \quad \text{normal}=a.$$

(24.) If an ordinate NP in an ellipse be produced until it meets the tangent, drawn from the extremity of the latus rectum, in T ; prove that the distance of P from the focus is equal to the distance of T from the axis of abscissæ.

(25.) In the ellipse, if it be assumed that $x=a\cos t$; prove that the equation to the tangent will be

$$bx\cos t+ay\sin t=ab.$$

(26.) Find the locus of the intersection of pairs of tangents to an ellipse, the tangents always intersecting each other at right angles. $\qquad x^2+y^2=a^2+b^2.$

(27.) $y^2 = \dfrac{x^3}{2a-x}$ being the equation to the cissoid of Diocles, find the equation to the tangent, and show that there is an asymptote which cuts the diameter at its extremity at right-angles.

Equation to tan. $y_1 = \left\{\dfrac{x}{(2a-x)^3}\right\}^{\frac{1}{2}} \cdot \{(3a-x)x_1 - ax\}$.

(28.) Prove that half the minor axis of an ellipse is a mean proportional between the normal and the perpendicular from the centre upon the tangent.

(29.) In the logarithmic curve, whose equation is $y=a^x$, show that the subtangent is equal to the modulus of the system whose base is a.

(30.) Prove that the curve whose subnormal is constant is a parabola.

(31.) In the hyperbola, whose equation is $y^2 = \dfrac{b^2}{a^2}(2ax+x^2)$, show that $y = \pm\dfrac{b}{a}(x+a)$ is the equation to two asymptotes passing through the centre and equally inclined to the axis of x.

(32.) Draw the rectilinear asymptotes of the curve defined by $y^4 + x^3 y = a^2 x^2$, and determine the form of the curve at the origin.

(33.) Let $x^3 - y^3 + ax^2 = 0$ be the equation to a curve; show that the equation to the asymptote is $y = x + \dfrac{a}{3}$.

(34.) If $ay^3 = bx^3 - c^2 xy$ be the equation to a curve; show that $y = \left(\dfrac{b}{a}\right)^{\frac{1}{3}} \cdot \left(x - \dfrac{c^2}{3a^{\frac{1}{3}}b^{\frac{2}{3}}}\right)$ is the equation to the asymptote.

(35.) In the common parabola, whose equation is $y^2 = 4ax$, find that point at which the angle, made by a straight line from the vertex with the curve, is a maximum.

$$x = 2a.$$

(36.) A rectangular hyperbola, and a circle whose radius is $2a$, have the same centre; find the angle of intersection of the two curves.

$$\text{Angle} = \tan^{-1}\frac{\sqrt{15}}{4}.$$

(37.) Find that point in an ellipse at which the angle contained between the normal and the line drawn to the centre is a maximum.

(38.) Determine the angle at which the curve, called the lemniscata of Bernouilli, whose equation is $(y^2 + x^2)^2 = 2a^2(x^2 - y^2)$, cuts the axis of x.

(39.) If A be the vertex, P and Q corresponding points in the cycloid and its generating circle, prove that the tangent at P is parallel to the chord AQ.

(40.) The centre of an ellipse is the vertex of a parabola, the axis of the parabola intersects the axis of the ellipse at an angle of 90°, and the curves also intersect each other at right angles; show that major axis : minor axis :: $\sqrt{2}$: 1.

(41.) If $y^2 = mx + nx^2$, show that an asymptote cuts the axes at points indicated by $x = -\dfrac{m}{2n}$ and $y = \dfrac{m}{2n^{\frac{1}{2}}}$.

(42.) Show that the locus of the intersection of tangents to the rectangular hyperbola and perpendiculars upon them from the centre is the lemniscata.

(43.) Draw the asymptotes of the curve $y^2 = \dfrac{(x+a)^5}{(x-a)^3}$, and determine the distance of its minimum ordinate from the origin.

(44.) Find that tangent to a given curve which cuts off from the co-ordinate axes the greatest area.

$$x_0 = 2x, \ y_0 = 2y.$$

(45.) Draw a tangent to the curve, whose equation is $y=ax^{\frac{m-1}{m}}$, and show that the tangent always cuts from the axis of y a portion equal to an mth part of the ordinate at the point of contact.

(46.) If $y^3+x^3-3x^2=0$, show that $y=-x+1$ is the equation to the asymptote, and that the maximum ordinate is at the point indicated by $x=2$.

(47.) If C be the centre of an ellipse, and NP any ordinate, and if in NP a point Q be so taken that its distance from C shall be equal to NP; show that the locus of Q is an ellipse whose major axis is the minor axis of the given ellipse.

(48.) Draw a tangent to the curve whose equation is $y=\dfrac{x^3}{a^2+x^2}$, and determine whether the curve has an asymptote.

(49.) ABD is a semicircle, centre C and diameter AD; EF is a chord parallel to AD, CQR a radius cutting EF in Q; QR is bisected in P. Find the locus of P.
$$ay=(2y-b)(x^2+y^2)^{\frac{1}{2}}.$$

(50.) Show that the curve, whose equation is x^3+aby $-axy=0$, has a rectilinear asymptote at the distance b from the origin, and also a parabolic asymptote, whose equation is $ay-\dfrac{3}{4}b^2=\left(x-\dfrac{1}{2}b\right)^2$, the latus rectum of the parabola being a, and its axis parallel to the axis of y.

(51.) BAC is a triangle, right-angled at A; a straight rod moves through the fixed point C, while one end slides down the line BA : show that the curve described by the other end is a conchoid whose equation is $x^2y^2=(x-b)^2(a^2-x^2)$, and determine its subtangent.

CHAPTER XIII.

POLAR CO-ORDINATES. SPIRALS.

If $r=f(\theta)$, or $p=f(r)$, and $u=\dfrac{1}{r}$; then

Tangent of angle (ϕ) contained by radius vector (r) and a tangent to the curve, is $\tan SPY = r\dfrac{d\theta}{dr} = -u\dfrac{d\theta}{du}$.

Perpendicular on tangent,

$$SY = p = \frac{r^2}{\sqrt{r^2+\dfrac{dr^2}{d\theta^2}}}.$$

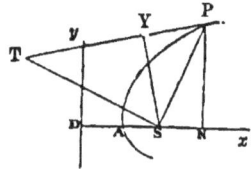

Subtangent $ST = r^2\dfrac{d\theta}{dr}$.

$$\frac{1}{p^2} = u^2 + \frac{du^2}{d\theta^2}. \qquad \frac{d\theta}{dr} = \frac{p}{r\sqrt{r^2-p^2}}.$$

If $A=$ area ANP, $\qquad \dfrac{dA}{d\theta} = \dfrac{1}{2}r^2$.

Ex. (1.) Find the polar equation to the common parabola.
$$SP=r, \qquad \angle ASP=\theta.$$
$$r = DN = 2AS+SN = 2a+r\cos PSN = 2a - r\cos\theta.$$

$$\therefore r+r\cos\theta = 2a, \qquad \therefore r = \frac{2a}{1+\cos\theta} = \frac{a}{\cos^2\dfrac{\theta}{2}}.$$

(2.) The equation to the spiral of Archimedes is $r=a\theta$; find the angle between the radius vector and tangent, and the subtangent.

$$\frac{dr}{d\theta} = a, \qquad \therefore SPY = \tan^{-1}r\frac{d\theta}{dr} = \tan^{-1}r\frac{1}{a} = \tan^{-1}\theta.$$

$$\text{Subtangent } ST = r^2\frac{d\theta}{dr} = \frac{r^2}{a}.$$

(3.) If $r=a\,(1+\cos\theta)$, find the equation between p and r.

$$\frac{1}{u}=a+a\cos\theta. \qquad \therefore u=\frac{1}{a+a\cos\theta}. \qquad \frac{du}{d\theta}=\frac{a\sin\theta}{(a+a\cos\theta)^2}.$$

$$\left(\frac{du}{d\theta}\right)^2=\frac{a^2\sin^2\theta}{(a+a\cos\theta)^4}=\frac{a^2(1-\cos^2\theta)}{(a+a\cos\theta)^4}.$$

But $a\cos\theta=\dfrac{1}{u}-a,$ $\qquad\qquad a^2\cos^2\theta=\dfrac{1}{u^2}-\dfrac{2a}{u}+a^2,$

$$\therefore a^2-a^2\cos^2\theta=\frac{2a}{u}-\frac{1}{u^2}, \qquad a+a\cos\theta=\frac{1}{u}.$$

$$\therefore u^2+\left(\frac{du}{d\theta}\right)^2=u^2+\frac{\dfrac{2a}{u}-\dfrac{1}{u^2}}{\dfrac{1}{u^4}}=u^2+2au^3-u^2=\frac{2a}{r^3}.$$

$$\therefore \frac{1}{p^2}=\frac{2a}{r^3}, \qquad p^2=\frac{r^3}{2a}.$$

(4.) The tangents at the vertex and extremity of the latus rectum of a conic section intersect; prove that the distance of the point of intersection from the vertex is equal to the distance of the focus from the vertex.

Let A be the vertex, S the focus, and T the point of intersection.

The equation $y=\dfrac{b}{a}\sqrt{2ax\mp x^2}$ will, by using the negative sign, comprehend all the conic sections excepting the hyperbola; and, by using the positive sign, it is the equation to that curve.

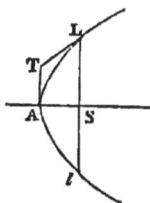

Also $y_,-y=\dfrac{dy}{dx}(x_,-x)$ is the equation to the tangent.

Differentiating the assumed equation, $\dfrac{dy}{dx}=\dfrac{b}{a}\cdot\dfrac{a\mp x}{\sqrt{2ax\mp x^2}};$

and substituting the values of y and $\dfrac{dy}{dx}$ in the equation to the tangent, we have

$$y_{,}=\frac{b}{a}\sqrt{2\,ax\mp x^2}+\frac{b}{a}\frac{a\mp x}{\sqrt{2\,ax\mp x^2}}\cdot(x_{,}-x).$$

But at the origin $x_{,}=0$, and $x=AS=m$ suppose. Then

$$y_{,}=\frac{b}{a}\sqrt{2\,am\mp m^2}-\frac{bm}{a}\frac{a\mp m}{\sqrt{2\,am\mp m^2}}$$

$$=\frac{b}{a}\left(\frac{2\,am\mp m^2-am\pm m^2}{\sqrt{2\,am\mp m^2}}\right)=\frac{bm}{\sqrt{2\,am\mp m^2}}\cdot$$

Now $a^2=b^2+(a\mp m)^2$, by a property of the curve,

$\therefore\ 2\,am\mp m^2=b^2$, $\qquad\therefore\ TA=y_{,}=m=SA.$

(5.) In the ellipse, if p be the perpendicular from the centre on the tangent, and r be the distance of the point in the curve from the centre, prove that $p^2=\dfrac{a^2b^2}{a^2+b^2-r^2}\cdot$

Perp^r $CP=r$, $\angle PCN=\theta$, then $x=r\cos\theta$, $y=r\sin\theta$;

$$\frac{x^2}{a^2}+\frac{y^2}{b^2}=1, \qquad \text{equation to the ellipse.}$$

$$\therefore\ \frac{r^2\cos^2\theta}{a^2}+\frac{r^2\sin^2\vartheta}{b^2}=r^2\left(\frac{\cos^2\theta}{a^2}+\frac{\sin^2\theta}{b^2}\right)=1.$$

$$\therefore\ r^2=\frac{a^2b^2}{b^2\cos^2\theta+a^2\sin^2\theta}=\frac{a^2b^2}{a^2(1-e^2)\cos^2\theta+a^2\sin^2\theta},$$

$$\text{where } 1-e^2=\frac{b^2}{a^2}\cdot$$

$$\therefore\ r^2=\frac{a^2b^2}{a^2-a^2e^2\cos^2\theta}=\frac{b^2}{1-e^2\cos^2\theta}\cdot$$

$$\therefore\ u^2=\frac{1}{b^2}\cdot(1-e^2\cos^2\theta), \qquad 2u\frac{du}{d\theta}=\frac{1}{b^2}\{-2e^2\cos\theta(-\sin\theta)\}\cdot$$

$$\frac{du}{d\theta}=\frac{1}{b^2}\{e^2\cos\theta\sin\theta\}\cdot\frac{1}{u}=\frac{1}{b}\cdot\frac{e^2\cos\theta\sin\theta}{\sqrt{1-e^2\cos^2\theta}}\cdot$$

$$u^2+\left(\frac{du}{d\theta}\right)^2=\frac{1-e^2\cos^2\theta}{b^2}+\frac{e^4\cos^2\theta\,(1-\cos^2\theta)}{b^2(1-e^2\cos^2\theta)}$$

$$=\frac{1-2e^2\cos^2\theta+e^4\cos^4\theta+e^4\cos^2\theta-e^4\cos^4\theta}{b^2(1-e^2\cos^2\theta)}$$

$$=\frac{1-2e^2\cos^2\theta+e^4\cos^2\theta}{b^2(1-e^2\cos^2\theta)}.$$

But $\dfrac{1}{r^2}=\dfrac{1-e^2\cos^2\theta}{b^2}$, $\therefore e^2\cos^2\theta=1-\dfrac{b^2}{r^2}$, $e^4\cos^2\theta=e^2-\dfrac{b^2e^2}{r^2}$.

$$\therefore\frac{1}{p^2}=\frac{1-2+\dfrac{2b^2}{r^2}+e^2-\dfrac{b^2e^2}{r^2}}{b^2\cdot\dfrac{b^2}{r^2}}=\frac{e^2-1-\dfrac{b^2}{r^2}(e^2-2)}{\dfrac{b^4}{r^2}}$$

$$=\frac{-\dfrac{b^2}{a^2}-\dfrac{b^2}{r^2}\left(-1-\dfrac{b^2}{a^2}\right)}{\dfrac{b^4}{r^2}}=\frac{-\dfrac{1}{a^2}+\dfrac{1}{r^2}\left(1+\dfrac{b^2}{a^2}\right)}{\dfrac{b^2}{r^2}}$$

$$=\frac{-r^2+a^2\left(1+\dfrac{b^2}{a^2}\right)}{a^2b^2}\qquad\therefore p^2=\frac{a^2b^2}{a^2+b^2-r^2}.$$

(6.) In the ellipse, if A, be the origin, the equation is $y^2=\dfrac{b^2}{a^2}(2ax-x^2)$: let S be the pole, $\angle A_{,}SP=\theta$, and $SP=r$; show that the equation referred to polar co-ordinates is

$$r=\frac{a(1-e^2)}{1+e\cos\theta}.$$

(7.) The equation to a curve being $y=(x^m+ax^{m-1})^{\frac{1}{m}}$; determine the polar equation, and show that an asymptote cuts the axis of abscissæ at an angle of $45°$, and at a distance $=-\dfrac{a}{m}$ from the origin of co-ordinates.

(8.) In the hyperbola, if S be the pole, the polar equation will be $r=\dfrac{a\,(e^2-1)}{1+e\cos\theta}$; if the centre be the pole, the

polar equation will be $r=\dfrac{b}{\sqrt{e^2\cos^2\theta-1}}$.

(9.) Show that the polar equation to the lemniscata of Bernouilli is $r^2=2\,a^2\cos 2\,\theta$, and that $p=\pm\dfrac{r^3}{2\,a^2}$.

(10.) Show that the polar equation to the conchoid of Nicomedes is $r=a+\dfrac{b}{\cos\theta}$, the equation between rectangular co-ordinates being $x^2y^2=(a+x)^2\,(b^2-x^2)$.

(11.) Show that the equation $r=\dfrac{a\theta}{\theta\pm\pi}$ represents two polar curves, one having an exterior and the other an interior asymptotic circle, and exhibit the general form of the two spirals.

(12.) The polar equation to the cissoid of Diocles is $r=2\,a\tan\theta\sin\theta$. Prove this.

(13.) The equation to the lituus is $r^2=\dfrac{a^2}{\theta}$; show that the subtangent $=2\,a\sqrt{\theta}$.

(14.) In the cardioid $r=a\,(1-\cos\theta)$, and if $r_,$ be a radius in the direction of r produced backwards, $r_,=a\,(1+\cos\theta)$: show that $2\phi=\theta$.

(15.) If the polar equation to a hyperbola, referred to its focus, be $r=\dfrac{a\,(e^2-1)}{1+e\cos\theta}$; show that there are two asymptotes intersecting the axis of x at a distance ae from the origin, at angles whose tangents are $+\dfrac{b}{a}$ and $-\dfrac{b}{a}$ respectively.

(16.) If $\theta=\dfrac{1}{\sqrt{2ar-r^2}}$ be the equation to a spiral; show that a circle whose radius is $2a$ is an asymptote to the spiral.

(17.) If $\theta = \dfrac{a^n}{r^n}$, and $na^n = b^n$; show that the equation between the radius vector and perpendicular on tangent is

$$p = \frac{b^n r}{\sqrt{b^{2n} + r^{2n}}}.$$

CHAPTER XIV.

SINGULAR POINTS.   TRACING OF CURVES.

A curve is convex or concave to the axis according as y and $\dfrac{d^2y}{dx^2}$ have the *same* or *opposite* signs.

To determine whether there be a point of contrary flexure, we put $\dfrac{d^2y}{dx^2} = 0$ or ∞; and if a be one of the values of x so found, we substitute successively $a+h$ and $a-h$ for x in $\dfrac{d^2y}{dx^2}$; then if $\dfrac{d^2y}{dx^2}$ have opposite signs, there will be a point of contrary flexure denoted by $x=a$.

At a point of contrary flexure in polar curves $\dfrac{dp}{dr} = 0$.

If any values of x and y make $\dfrac{dy}{dx} = \dfrac{0}{0}$, this circumstance generally indicates a multiple point.

For a true double point
$$\left(\frac{d^2u}{dx\,dy}\right)^2 - \left(\frac{d^2u}{dx^2}\right)\left(\frac{d^2u}{dy^2}\right) > 0.$$

For a point of osculation
$$\left(\frac{d^2u}{dx\,dy}\right)^2 - \left(\frac{d^2u}{dx^2}\right)\left(\frac{d^2u}{dy^2}\right) = 0.$$

For a conjugate point
$$\left(\frac{d^2u}{dx\,dy}\right)^2 - \left(\frac{d^2u}{dx^2}\right)\left(\frac{d^2u}{dy^2}\right) < 0.$$

At a cusp, if $x=a$, $\dfrac{dy}{dx}$ has but one value; and, substituting successively $a+h$ and $a-h$ for x, $\dfrac{d^2y}{dx^2}$ has two values.

For the ceratoid or cusp of the first species, the values of $\dfrac{d^2y}{dx^2}$ have *opposite* signs.

For the ramphoid or cusp of the second species, the values of $\dfrac{d^2y}{dx^2}$ have the *same* sign.

Ex. (1.) If the equation to a curve be $y=\dfrac{1}{a}\sqrt{x^5+cx^4}$; show that the origin is a point of osculation, ascertain if there be any maximum ordinate, and determine the general form of the curve.

It is obvious that, by giving x successive positive values from 0 to ∞, y will have successive positive and negative values from 0 to ∞, consequently there are two similar branches extending from the origin to infinity, one branch on each side of the axis of x to the right of the axis of y.

Now $\dfrac{dy}{dx}=\dfrac{1}{a}\cdot\dfrac{5x^4+4cx^3}{2\sqrt{x^5+cx^4}}=\dfrac{x}{2a}\cdot\dfrac{5x+4c}{\sqrt{x+c}}=0$ when $x=0$,

and $\because$ when $x=0$, y also $=0$, and $\dfrac{dy}{dx}$ has two values, one positive and the other negative, each $=0$, therefore the axis of x is

a common tangent to the two infinite branches at the origin; hence the origin is a point of osculation.

Again $\because$ $y=\dfrac{x^2}{a}\sqrt{x+c}$; when $x=-c$, $y=0$, and while x takes successive negative values from 0 to $-c$, y will take successive positive and negative values from 0 to 0 again,

K 2

and therefore to the left of the axis of y there is a loop or nodus.

And $\because \dfrac{dy}{dx}=\dfrac{x}{2a}\cdot\dfrac{5x+4c}{\sqrt{x+c}}=0,\quad \therefore 5x+4c=0,$ and $x=-\dfrac{4}{5}c$

determines the position of the maximum double ordinate;

and $\because \dfrac{dy}{dx}=\tan\theta=\infty$ when $x=-c,$ the tangent at this point

intersects the axis of x at right-angles.

Take $AB=c,$ and draw the tangent $TBt\perp AB,$ take

$AN=\dfrac{4}{5}c,$ and draw the double ordinate $PNp=\dfrac{32}{a}\left(\dfrac{c}{5}\right)^{\frac{5}{2}},$

which is the value of $2y$ corresponding to $x=-\dfrac{4}{5}c$; the loop will pass through $A, P, B, p.$

(2.) Trace the curve, whose equation is $y=\dfrac{\sqrt{x}}{\sqrt{a}}(a\pm x);$

and show that there is an oval between $x=0$ and $x=a$; determine the position of the maximum double ordinate, and exhibit the form of the exterior branch.

Firstly, $y=\dfrac{\sqrt{x}}{\sqrt{a}}(a-x)=\sqrt{a}\sqrt{x}-\dfrac{x^{\frac{3}{2}}}{\sqrt{a}}.$

Let $x=0,\ \therefore y=0,$ 　Take $AB=a.$

$\quad x<a,\quad y$ is $\pm,$ 　Then, $\because$ while

$\quad x=a,\quad y=0,$ 　x increases

$\quad x>a,\quad y$ is impossible. 　from 0 to $a,$ y has positive

Putting $-x$ for $x,$ y is impossible. 　and negative values from

　　　　　　　　　　　　　0 to 0 again, $\therefore$ there is

a maximum ordinate somewhere between A and $B,$ and AB is the axis of an oval.

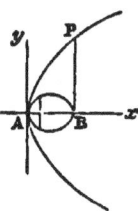

Now $\dfrac{dy}{dx}=\sqrt{a}\cdot\dfrac{1}{2\sqrt{x}}-\dfrac{\frac{3}{2}x^{\frac{1}{2}}}{\sqrt{a}}=\dfrac{\sqrt{a}}{2\sqrt{x}}-\dfrac{3}{2\sqrt{a}}\sqrt{x}=0,$

$$\therefore \frac{\sqrt{a}}{\sqrt{x}} = \frac{3\sqrt{x}}{\sqrt{a}}, \quad 3x = a, \quad \therefore x = \frac{a}{3} \text{ denotes the point where}$$

the maximum double ordinate cuts the axis of x.

Secondly, $y = \dfrac{\sqrt{x}}{\sqrt{a}}(a+x)$.

Let $x=0$, $\therefore y=0$,
$\quad x<a, \quad y$ is $\pm$,
$\quad x=a, \quad y=2a$,
$\quad x>a, \quad y$ is $\pm$,
$\quad x=\infty, \quad y=\infty$,
Putting $-x$ for x, y is impossible.

Draw $BP=2a$. Then, $\because$ while x increases from 0 to infinity, y has positive and negative values from 0 to infinity; there is a branch above and below the axis of x exterior to the oval.

No curve exists to the left of the origin.

(3.) $y^2(a^2+x^2) = x^2(a^2-x^2)$ is the equation to a curve; trace it, determine the angles at which it cuts the axis of x, and find its maximum ordinate.

$$y = x \cdot \sqrt{\frac{a^2-x^2}{a^2+x^2}}.$$

If $x=0$, then $y=0$
$\quad x<a, \quad y$ is possible $\pm$
$\quad x=a, \quad y=0$
$\quad x>a, \quad y$ is impossible.

Put $-x$ for x, then
$\quad$ if $x=0$, $y=0$
$\quad x<a$, y is possible $\mp$
$\quad x=a$, $y=0$
$\quad x>a$, y is impossible.

Take $AB=a$, $Ab=-a$, in the axis of x, and the curve will pass through the points A, B, b.

And $\because$ when $x>a$, y is impossible, the curve cannot extend beyond B, b.

Now $2y\dfrac{dy}{dx} = x^2 \cdot \dfrac{(a^2+x^2)(-2x)-(a^2-x^2)(2x)}{(a^2+x^2)^2} + \dfrac{a^2-x^2}{a^2+x^2} \cdot 2x.$

$$y\frac{dy}{dx} = \frac{-a^2x^3 - x^5 - a^2x^3 + x^5 + a^4x - x^5}{(a^2 + x^2)^2}$$

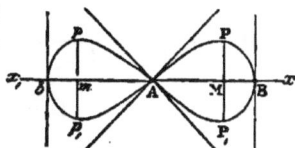

$$= \frac{a^4x - 2a^2x^3 - x^5}{(a^2 + x^2)^2},$$

$$\therefore \frac{dy}{dx} = \frac{1}{x} \cdot \frac{(a^2 + x^2)^{\frac{1}{2}}}{(a^2 - x^2)^{\frac{1}{2}}} \cdot \frac{x(a^4 - 2a^2x^2 - x^4)}{(a^2 + x^2)^2} = \frac{a^4 - 2a^2x^2 - x^4}{(a^2 + x^2)^{\frac{3}{2}}(a^2 - x^2)^{\frac{1}{2}}},$$

and putting $x = 0$ and $\pm a$ in this expression, we have

$$\tan\theta = \frac{dy}{dx} = \frac{a^4}{(a^2)^{\frac{3}{2}}(a^2)^{\frac{1}{2}}} = \frac{a^4}{a^3 \cdot a} = \pm 1 = \tan 45° \text{ or } \tan 135°.$$

$$= \frac{-2a^4}{(2a^2)^{\frac{3}{2}}(0)} = \infty = \tan 90°.$$

$\therefore$ the two tangents at the point A are inclined to the axis of x at $\angle$ s $= 45°$ and $135°$ respectively, and the tangents at B and b are $\perp$ to the axis of x: $\therefore$ the point A is a double point.

To find the greatest ordinate, $y = x \cdot \sqrt{\dfrac{a^2 - x^2}{a^2 + x^2}}$, a max.

$$\therefore \frac{dy}{dx} = \frac{a^4 - 2a^2x^2 - x^4}{(a^2 - x^2)^{\frac{1}{2}}(a^2 + x^2)^{\frac{3}{2}}} = 0, \qquad \therefore a^4 - 2a^2x^2 - x^4 = 0,$$

$$x^4 + 2a^2x^2 + a^4 = 2a^4, \quad x^2 + a^2 = a^2\sqrt{2}, \quad x = \pm a\sqrt{\sqrt{2} - 1}.$$

Hence the greatest ordinate cuts the axis of x at points denoted by $x = a\sqrt{\sqrt{2} - 1}$ and $-a\sqrt{\sqrt{2} - 1}$, and the length of this ordinate may be ascertained by substituting these values of x in the equation to the curve. Thus

$$y = a\sqrt{\sqrt{2} - 1} \cdot \sqrt{\frac{a^2 - a^2\sqrt{2} + a^2}{a^2 + a^2\sqrt{2} - a^2}}$$

$$= a\sqrt{\sqrt{2} - 1} \cdot \sqrt{\frac{2a^2 - \sqrt{2} \cdot a^2}{\sqrt{2} \cdot a^2}}$$

$$= a\sqrt{\sqrt{2} - 1} \cdot \sqrt{\sqrt{2} - 1} = a(\sqrt{2} - 1)$$

$$= MP, \ MP_{,,} \ mp, \ mp_{,}.$$

(4.) If $y=\dfrac{a^2x}{a^2+x^2}$, show that there are points of contrary flexure when $x=0$ and $a\sqrt{3}$, that the curve cuts the axis of x at an angle of $45°$, that the axis of x is an asymptote to the two infinite branches, and that there are maximum ordinates when $x=+a$ and $-a$.

Let $x=0$, $\therefore y=0$　　　Put $-x$ for x, then $y=\dfrac{-a^2x}{a^2+x^2}$.

$x<a$,　　y is $+$　　　Let $x=0$, $\therefore y=0$

$x=a$,　　$y=\dfrac{a}{2}$　　　　$x<a$,　　y is $-$

$x>a$,　　y is $+$　　　$x=a$,　　$y=-\dfrac{a}{2}$

$x=\infty$,　$y=0$.　　　　$x>a$,　　y is $-$

　　　　　　　　　　　$x=\infty$,　$y=0$.

Take $AB=a$, $Ab=-a$, and draw the ordinates BQ, bq, equal to $\dfrac{a}{2}$ and $-\dfrac{a}{2}$

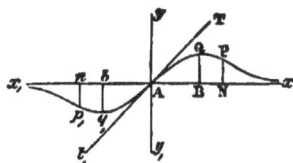

respectively, the curve will pass through the points A, Q, q,, its right-hand branch being above the axis of x, and its left-hand branch below it, the two branches meeting that axis again only at an infinite distance from the origin A. $\therefore$ the axis of x is an asymptote to the two infinite branches.

Now $\dfrac{dy}{dx}=\dfrac{(a^2+x^2)\,a^2-a^2x\cdot 2x}{(a^2+x^2)^2}=\dfrac{a^2\,(a^2-x^2)}{(a^2+x^2)^2}$.

$\dfrac{d^2y}{dx^2}=\dfrac{(a^2+x^2)^2\cdot(-2a^2x)-a^2(a^2-x^2)\cdot 2\,(a^2+x^2)\cdot 2x}{(a^2+x^2)^4}$

$\qquad=\dfrac{2a^2x\,(x^2-3a^2)}{(a^2+x^2)^3}=0$, if $x=a\sqrt{3}$ or 0.

Substituting $a\sqrt{3}-h$, $a\sqrt{3}+h$ respectively for x, we have

$$\frac{d^2y}{dx^2} = \frac{2a^2(a\sqrt{3}-h)\{(a\sqrt{3}-h)^2-3a^2\}}{\{a^2+(a\sqrt{3}-h)^2\}^3}$$

$$= \frac{-2a^2h(a\sqrt{3}-h)(2a\sqrt{3}-h)}{\{a^2+(a\sqrt{3}-h)^2\}^3},$$

which is *negative*, since $h < a\sqrt{3}$;

$$\frac{d^2y}{dx^2} = \frac{2a^2h(a\sqrt{3}+h)(2a\sqrt{3}+h)}{\{a^2+(a\sqrt{3}+h)^2\}^3}, \text{ which is } positive.$$

Hence $x = a\sqrt{3}$ indicates a point of contrary flexure ; and, substituting this value of x in the given equation, we have $y = \frac{a\sqrt{3}}{4}$. Take $AN = a\sqrt{3}$, and draw $NP = \frac{a\sqrt{3}}{4}$, when P will be a point of contrary flexure.

Also substituting $0-h$, $0+h$ respectively for x,

$$\frac{d^2y}{dx^2} = \frac{-2a^2h(h^2-3a^2)}{(a^2+h^2)^3}, \qquad \frac{d^2y}{dx^2} = \frac{2a^2h(h^2-3a^2)}{(a^2+h^2)^3},$$

one *positive*, the other *negative*. $\therefore$ the origin A is also a point of contrary flexure.

Hence also, y being positive and $\frac{d^2y}{dx^2}$ to the left of NP negative, the curve from A to P is concave to the axis of x, and consequently beyond P it is convex.

Again $\because$ as x increases y at first increases and afterwards decreases, having various finite values between its primary value 0 and its ultimate value 0, there will be a maximum ordinate somewhere on each side of the origin.

$$\therefore \frac{dy}{dx} = \frac{a^2(a^2-x^2)}{(a^2+x^2)^2} = 0, \qquad \therefore a^2-x^2 = 0, \qquad x = \pm a.$$

But when $x = \pm a$, $y = \pm\frac{a}{2}$. Draw $BQ = \frac{a}{2}$, it will be a maximum ordinate.

By substituting 0 for x in $\dfrac{dy}{dx}$ we have

$\tan\theta = \dfrac{a^4}{a^4} = 1 = \tan 45°$. $\therefore$ the curve cuts the axis of x at the origin A at an $\angle$ of 45°.

(5.) If $y = x \cdot \sqrt{\dfrac{a^2 + x^2}{a^2 - x^2}}$; show that the branches of the curve pass through the origin, and are contained between two asymptotes perpendicular to the axis of x.

Let $x = 0$, $\therefore y = 0$ Put $-x$ for x, then

$x < a$,	y is possible $\pm$	if $x = 0$,	$y = 0$
$x = a$,	$y = \infty$	$x < a$,	y is $\mp$
$x > a$,	y is impossible.	$x = a$,	$y = -\infty$
		$x > a$,	y is impossible.

Take $AB = a$, $Ab = -a$; then, since at the origin A the ordinate is 0, and then as x increases the ordinates increase until $x = a$, when an infinite ordinate passes through B; and, since the values of y are both positive and negative, a branch extends on each side of the axis of x.

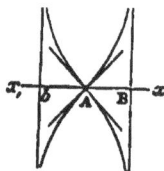

Also, since when x is negative, the ordinates take values exactly corresponding to those when x is positive, the curve has similar branches to the left of the origin.

Again $\dfrac{dy}{dx} = \dfrac{a^4 + 2\,a^2 x^2 - x^4}{(a^2 - x^2)^{\frac{3}{2}}\,(a^2 + x^2)^{\frac{1}{2}}}$: and, putting $x = 0$ and $\pm a$ in this expression, we have $\tan\theta = \dfrac{dy}{dx} = \pm 1$ and ∞.

$\therefore \tan\theta = 1 = \tan 45°$, $\tan\theta = -1 = \tan 135°$, $\tan\theta = \infty = \tan 90°$.

Hence a tangent to the curve cuts the axis of x in the origin A at an angle of 45°, another through the same point at an angle of 135° : and at B a tangent to the curve is $\perp$ the axis of x, and is coincident with the infinite ordinate. This tangent is consequently an asymptote, the branches of the curve do not extend beyond it, and they are convex to the axis of x.

(6.) If $(y-b)^2=(x-a)^5$; show that there is a ceratoid cusp when $x=a$, and that the tangent at that point is parallel to the axis of x.

If $x=a$, $y=b$.   Take $AB=a$, $BP=b$, then P is the point.

$$\text{Now } y-b=\pm(x-a)^{\frac{5}{2}}, \qquad \therefore \frac{dy}{dx}=\pm\frac{5}{2}(x-a)^{\frac{3}{2}}=0$$

when $x=a$; $\therefore \tan\theta=0$,   and the tangent to the curve at the point denoted by $x=a$ is $\parallel$ to the axis of x.

$$\text{Again } \frac{d^2y}{dx^2}=\pm\frac{15}{4}(x-a)^{\frac{1}{2}}=0 \text{ when } x=a ;$$

and, putting $a+h$, $a-h$ successively for x,

$$\frac{d^2y}{dx^2}=\pm\frac{15}{4}\sqrt{h}, \text{ which has two values, one } +, \text{ another } -.$$

$$\frac{d^2y}{dx^2}=\pm\frac{15}{4}\sqrt{-h}, \text{ which is imaginary :}$$

and since if $x=a$, $\dfrac{dy}{dx}=0$, $\dfrac{d^2y}{dx^2}=0$; and if $x=a-h$, they are both impossible $\therefore$ the curve cannot extend to the left of P: also $\because$ if $x=a+h$, $\dfrac{d^2y}{dx^2}$ has

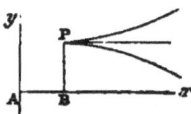

two values, one positive and the other negative, $\therefore$ at the point P there is a cusp of the first species.

(7.) Show that the curve, whose equation is $r=\dfrac{a\theta^2}{\theta^2-1}$, has

a point of inflection when $r=\dfrac{3a}{2}$, and rectilinear and circular asymptotes.

$$r\vartheta^2 - r = a\theta^2, \qquad (r-a)\,\theta^2 = r, \qquad \therefore \theta = \sqrt{\dfrac{r}{r-a}}\,.$$

$$\frac{d\theta}{dr} = \frac{\dfrac{r-a-r}{(r-a)^2}}{2\sqrt{\dfrac{r}{r-a}}} = -\frac{a}{2\,r^{\frac12}\,(r-a)^{\frac32}}. \qquad \text{But } \frac{d\theta}{dr} = \frac{p}{r\,(r^2-p^2)^{\frac12}},$$

$$\therefore \frac{p}{r\,(r^2-p^2)^{\frac12}} = -\frac{a}{2\,r^{\frac12}\,(r-a)^{\frac32}}, \qquad \frac{p^2}{r^2-p^2} = \frac{a^2 r}{4\,(r-a)^3},$$

$$\frac{r^2-p^2}{p^2} = \frac{r^2}{p^2} - 1 = \frac{4\,(r-a)^3}{a^2 r}, \qquad \frac{r^2}{p^2} = \frac{4\,(r-a)^3}{a^2 r} + 1,$$

$$\frac{1}{p^2} = \frac{4\,(r-a)^3 + a^2 r}{a^2 r^3}, \qquad \therefore p = \frac{a r^{\frac12}}{\sqrt{4\,(r-a)^3 + a^2 r}},$$

$$\frac{dp}{dr} = \frac{\dfrac{3}{2} a r^{\frac12}\sqrt{4\,(r-a)^3 + a^2 r} - a r^{\frac32}\cdot\dfrac{12\,(r-a)^2 + a^2}{2\sqrt{4\,(r-a)^3 + a^2 r}}}{4\,(r-a)^3 + a^2 r} = 0.$$

$$\therefore 3\,\{4\,(r-a)^3 + a^2 r\} - 12\,r\,(r-a)^2 - a^2 r = 0,$$

$$r^2 - \frac{13\,a}{6}\,r = -a^2, \qquad \therefore r = \frac{3\,a}{2}\,.$$

Hence there is a point of contrary flexure, when $r = \dfrac{3}{2}\,a$.

Again $\qquad \dfrac{1}{r} = \dfrac{\theta^2 - 1}{a\theta^2}$. Let r become infinitely great, then

$$\frac{1}{r} = \frac{1}{\infty} = 0, \qquad \therefore \theta^2 - 1 = 0, \qquad \theta = \pm 1.$$

$$\frac{d\theta}{dr} = -\frac{a}{2\,r^{\frac12}\,(r-a)^{\frac32}}, \qquad r^2\frac{d\theta}{dr} = -\frac{a r^{\frac32}}{2\,(r-a)^{\frac32}} = \mp\frac{a}{2}\left(\frac{r}{r-a}\right)^{\frac32};$$

and, when r becomes infinitely great,

$$\frac{r}{r-a} = \frac{1}{1 - \dfrac{a}{r}} = \frac{1}{1 - \dfrac{a}{\infty}} = \frac{1}{1-0} = 1,$$

$$\therefore \text{Subtangent } ST = r^2\frac{d\theta}{dr} = \mp\frac{a}{2}:$$

L

and, since ST remains finite while SP is infinite, a tangent may be drawn which will touch the curve at a point infinitely distant from the origin; this tangent is therefore a rectilinear asymptote: and $\because$ θ and ST have each two values, $\therefore$ there are two rectilinear asymptotes.

Again, let $r=a$, $\therefore \dfrac{1}{a}=\dfrac{\theta^2-1}{a\theta^2}$, $1=\dfrac{\theta^2-1}{\theta^2}=1-\dfrac{1}{\theta^2}$,

$\therefore \dfrac{1}{\theta^2}=0$, $\therefore \theta=\infty$ when $r=a$.

Also $\theta=\sqrt{\dfrac{r}{r-a}}$, which is impossible when $r<a$.

Hence $\because$ $r=a$ makes θ infinite, and $r<a$ makes θ impossible, there is an asymptotic $\odot$, radius $=a$, within the curve.

In the logarithmic and many other spirals the curve makes an infinite number of revolutions about the pole before reaching it; hence the pole may, in such instances, be considered as an indefinitely small asymptotic circle, that is, an asymptotic circle whose radius $=0$.

The equation to the logarithmic spiral is $r=a^\theta$, or $r=ae^{m\theta}$, or $r=ce^{\frac{\theta}{a}}$; r increasing in a geometric ratio, while θ increases in an arithmetic ratio; the radii including equal angles are proportional. Its evolute and involute are similar to the original spiral.

(8.) Trace the curve whose equation is $r=a\,(2\cos\theta\pm1)$.

Let $\theta=0$, $\therefore r=a\,(2+1)=3a$,

$\quad\theta=30°$, $r=a\,(\sqrt{3}+1)$, which is $<3a$,

$\quad\theta=60$, $r=a\,(1+1)=2a$,

$\quad\theta=90$, $r=a\,(0+1)=a$,

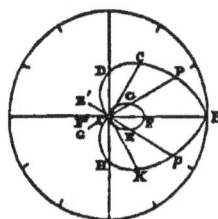

Let $\theta=120$, $\therefore \cos\theta = -\cos60 = -\dfrac{1}{2}$, $r=a\,(-1+1)=0$,

$\theta=150$, $\cos\theta = -\cos30 = -\dfrac{\sqrt{3}}{2}$, $r=a(-\sqrt{3}+1)$,

which is $< a$,

$\theta=180$, $\cos\theta=-1$, $r=a\,(-2+1)=-a$,

$\theta=210$, $\cos\theta = -\cos30 = -\dfrac{\sqrt{3}}{2}$, $r=a\,(-\sqrt{3}+1)$,

which is $< -a$,

$\theta=240$, $\cos\theta = -\cos60 = -\dfrac{1}{2}$, $r=0$,

$\theta=270$, $\cos\theta=0$, $r=a\,(0+1)=a$,

$\theta=300$, $\cos\theta=\cos60$, $r=a\,(1+1)=2a$,

$\theta=330$, $\cos\theta=\cos30$, $r=a\,(\sqrt{3}+1)$,

which is $> 2a$,

$\theta=360$, $\cos\theta=1$, $r=a\,(2+1)=3a$.

Divide the $\odot^{cc}$ of a $\odot$ into 12 equal parts, and draw radii through the points of division. Take $AB=3a$, AP, Ap each $=a\,(\sqrt{3}+1)$, AC, AK each $=2a$, AD, AH each $=a$.

Take AE', AG' each $=a(-\sqrt{3}+1)$, and $AF''=-a$. These three, being negative values of r, must be measured in an exactly opposite direction, as AE, AF, AG.

The curve, which is the trisectrix, will pass through the points B, P, C, D, A, H, K, p ; and the interior oval will pass through A, E, F, G.

Taking $r=a\,(2\cos\theta-1)$, a precisely similar curve is produced, but turned the contrary way.

Taking $-\theta$ for θ, the same curve is produced, $\because\ 2\cos(-\theta)=2\cos\theta$.

(9.) Show that the curve, whose equation is $(y^2+x^2)^3=4a^2x^2y^2$, has a quadruple point at the origin, and that there are four loops or ovals; namely, one in each quadrant.

Let the equation be transformed into one under polar coordinates, putting $x=r\cos\theta$, $y=r\sin\theta$.

$$(r^2\sin^2\theta+r^2\cos^2\theta)^3=4a^2r^2\sin^2\theta\,r^2\cos^2\theta, \qquad r^6=4a^2r^4\sin^2\theta\cos^2\theta,$$

$$r^2=4a^2\sin^2\theta\cos^2\theta, \qquad r=2a\sin\theta\cos\theta. \qquad \therefore\ r=a\sin2\theta$$

1st quad. If $\theta=0$, $\quad r=0$,

$\theta=15°$, $\quad r=a\sin30=\dfrac{a}{2}$,

$\theta=30$, $\quad r=a\sin60=\dfrac{\sqrt{3}}{2}a$,

$\theta=45$, $\quad r=a\sin90=a$,

$\theta=60$, $\quad r=a\sin120=\dfrac{\sqrt{3}}{2}a$,

$\theta=75$, $\quad r=a\sin150=\dfrac{a}{2}$,

$\theta=90$, $\quad r=a\sin180=0$,

2nd quad. $\quad\theta=105$, $\quad r=a\sin210=-\dfrac{a}{2}$,

3rd quad. $\quad\theta=195$, $\quad r=a\sin390=\dfrac{a}{2}$,

4th quad. $\quad\theta=285$, $\quad r=a\sin570=-\dfrac{a}{2}$.

By putting $-\theta$ for θ the curve is reproduced.

Take the several values of r at the corresponding angles.

In the second and fourth quadrants, the values of r, being negative, must be measured in opposite directions.

Hence, there will be an oval whose axis $=a$ in each quadrant: and the origin is a quadruple point.

(10.) If $r=a\tan\theta$, show that the asymptotic subtangent is a, and that the curve is included between vertical asymptotes.

Let $\theta=0$, $\therefore r=0$,   Let $\theta=\pi+45$, $\therefore r=a$,

$\theta=45°$, $r=a$, $\theta=\dfrac{3\pi}{2}$, $r=\infty$,

$\theta=\dfrac{\pi}{2}$, $r=\infty$, $\theta=\dfrac{3\pi}{2}+45$, $r=-a$,

$\theta=135$, $r=-a$, $\theta=2\pi$, $r=0$.

$\theta=\pi$, $r=0$.

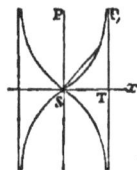

Take therefore $SB=a$ at an angle of $45°$ with the axis of x, the curve will pass from the origin S through B to infinity.

And $\because$ those lines are said to be $\parallel$ which coincide only at an infinite distance, and $\because$ the asymptote will ultimately coincide with the curve and consequently with SP when both are infinite, $\therefore$ the asymptote must be drawn $\parallel SP$.

There are similar branches in all the four quadrants.

Now $\dfrac{dr}{d\theta}=a\,(1+\tan^2\theta)$, $\dfrac{d\theta}{dr}=\dfrac{1}{a\,(1+\tan^2\theta)}$,

$r^2\cdot\dfrac{d\theta}{dr}=\dfrac{a^2\tan^2\theta}{a\,(1+\tan^2\theta)}=a\,\dfrac{\tan^2\theta}{\sec^2\theta}=a\cdot\dfrac{\infty}{\infty}=a$, when $\theta=\dfrac{\pi}{2}$.

$\therefore ST=r^2\dfrac{d\theta}{dr}=a$, the asymptotic subtangent.

Take $ST=a$, and draw $TP, \parallel SP$; $TP,$ produced is the asymptote. Hence, this curve is included between vertical asymptotes.

(11.) $x=a\,(1-\cos\theta)$, $y=a\theta$ are equations to the curve called the companion to the cycloid; find the points of contrary flexure.

Let BDQ be the generating circle, centre O, vertex D, radius $=a$, $DM=x$, $MP=y$, $\angle DOQ=\theta$.

Let $\theta=0$, $\therefore x=0$, $y=0$,

$\theta=\dfrac{\pi}{2}$, $x=a$, $y=\dfrac{\pi}{2}a$,

$\theta=\dfrac{\pi}{2}+\alpha$, $\therefore \cos\theta=-\sin\alpha$.

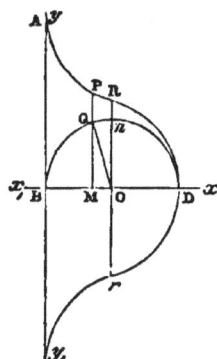

$\left.\begin{array}{l} x=a\,(1+\sin\alpha), \\[2mm] y=a\left(\dfrac{\pi}{2}+\alpha\right), \end{array}\right\}$ which increase as α increases.

Let $\alpha=\dfrac{\pi}{2}$, $\theta=\pi$, $\therefore \cos\theta=-1$, $-\cos\theta=1$,

$$x=a\,(1+1)=2\,a, \qquad y=\pi a.$$

Putting $-\theta$ for θ, a similar curve is produced on the other side of the axis of x.

$$\text{Now}\quad \frac{dy}{dx}=\frac{a}{(2\,ax-x^2)^{\frac{1}{2}}},$$

$$\frac{d^2y}{dx^2}=\frac{-\dfrac{1}{2}a\,(2\,ax-x^2)^{-\frac{1}{2}}(2a-2x)}{2\,ax-x^2}=\frac{-a\,(a-x)}{(2\,ax-x^2)^{\frac{3}{2}}}=0,\ \text{ if }x=a.$$

Substituting $a+h$, $a-h$ respectively for x in this expression, we have

$$\frac{d^2y}{dx^2}=\frac{-a\{a-(a+h)\}}{\{2a\,(a+h)-(a+h)^2\}^{\frac{3}{2}}}=\frac{-a\,(-h)}{(a^2-h^2)^{\frac{3}{2}}},\ \text{ which is positive,}$$

$$\frac{d^2y}{dx^2}=\frac{-a\{a-(a-h)\}}{\{2a\,(a-h)-(a-h)^2\}^{\frac{3}{2}}}=\frac{-ah}{(a^2-h^2)^{\frac{3}{2}}},\ \text{ which is negative;}$$

$\therefore$ there is a point of contrary flexure when $x=a$, $y=\dfrac{\pi}{2}a$.

$DO=a$.   Take $OR=\dfrac{\pi}{2}a$, $Or=-\dfrac{\pi}{2}a$, each $=$ arc Dn,

$BA=\pi a=$ arc DQB; the curve will pass through D, R, A, and R, r will be the points of contrary flexure.

(12.) Show that the curve $y^4 + 2axy^2 - ax^3 = 0$ has a triple point at the origin, and determine the position of the tangents.

$$4y^3p + 2a(x \cdot 2yp + y^2) - 3ax^2 = 0, \qquad \text{where } p = \frac{dy}{dx};$$

$$(4y^3 + 4axy)p = 3ax^2 - 2ay^2,$$

$$p = \frac{3ax^2 - 2ay^2}{4y^3 + 4axy} = \frac{0}{0}, \qquad \text{if } x = 0 \text{ and } y = 0.$$

∴ there may be a multiple point.

Differentiating numerator and denominator,

$$p = \frac{6ax - 4ayp}{12y^2p + 4axp + 4ay} = \frac{0}{0}, \qquad \text{if } x = 0 \text{ and } y = 0.$$

Differentiating as before,

$$p = \frac{6a - 4ayq - 4ap^2}{24yp^2 + 12y^2q + 4axq + 4ap + 4ap}, \qquad {}^{*} \qquad \text{where } \frac{d^2y}{dx^2} = q,$$

$$p = \frac{6a - 4ap^2}{8ap} = \frac{3 - 2p^2}{4p}, \qquad \text{if } x = 0 \text{ and } y = 0.$$

$$\therefore 4p^2 = 3 - 2p^2, \qquad \therefore p = \pm \frac{1}{\sqrt{2}}.$$

Also $p = \dfrac{3ax^2 - 2ay^2}{4y^3 + 4axy} = -\dfrac{2ay^2}{4y^3}, \qquad \text{if } x = 0,$

$$\therefore p = -\frac{a}{2y} = -\frac{a}{0} = -\infty, \qquad \text{if } y = 0,$$

∴ the origin is a triple point; and ∵ $\tan\theta = \dfrac{dy}{dx} = +\dfrac{1}{\sqrt{2}}$

and $= -\dfrac{1}{\sqrt{2}}$ and also $= \infty$, ∴ the tangents cut the axis

at $\angle s = \tan^{-1}\left(\dfrac{1}{\sqrt{2}}\right)$ and $\tan^{-1}\left(-\dfrac{1}{\sqrt{2}}\right)$, and at right-angles.

---

* These repeated differentiations are sometimes tedious: they may, however, in such cases as this, be simplified by considering p constant, as no error will arise from that assumption. Thus, instead of this equation, we should have had, by considering p in the previous one constant, $p = \dfrac{6a - 4ap^2}{24yp^2 + 4ap + 4ap}$, whence $p = \pm\dfrac{1}{\sqrt{2}}$ as above.

(13.) In the diameter AB of a circle take a point C, draw a chord AP and an ordinate PN, and CQ parallel to AP, meeting PN in Q: trace the curve which is the locus of Q.

$$AB=a, \quad AC=b, \quad AN=x, \quad NQ=y.$$

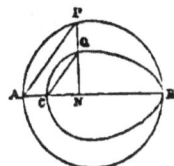

$$NP=\sqrt{ax-x^2}, \qquad \text{equation to } \odot,$$

$$CN \ : \ AN \ :: \ NQ \ : \ NP, \quad \text{or}$$

$$x-b \ : \ \dot{x} \ :: \ y \ : \ \sqrt{ax-x^2},$$

$$\therefore xy=(x-b)\sqrt{ax-x^2}, \qquad \therefore y=(x-b)\sqrt{\frac{a-x}{x}} \text{ is}$$

the equation to the curve which is the locus of Q.

Let $y=0$, $\therefore x=b$ and $=a$; let $x>a$, y is impossible. y has finite values positive and negative when $x>b$ and $<a$.

Hence the curve will pass through C, Q, B, and form an oval.

By the question no part of the curve can be to the left of C.

(14.) A rod PQ passes through a fixed point A; find the equation to the curve described by P when Q moves in the circumference of a circle of given radius, and trace the curve.

$PQ=R=$ length of rod, diameter of $\odot$ $BQ=a$, $AB=b$, qp position of rod when Q has moved along the arc Qq, $AN=x$, $Nq=y$; then $Nq^2=BN \cdot NQ$. Euc. iii. 35.

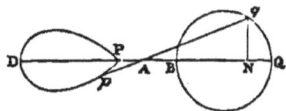

$$y^2=(x-b)\cdot(a+b-x)=(x-b)(c-x), \quad \text{if } c=a+b,$$
$$=-x^2+(b+c)x-bc.$$

Let $Aq=r$, $\angle A=\theta$, $\therefore y=r\sin\theta$, $x=r\cos\theta$,

$$r^2\sin^2\theta=-r^2\cos^2\theta+(b+c)r\cos\theta-bc,$$
$$r^2-(b+c)\cos\theta\cdot r=-bc,$$
$$\therefore r=\frac{1}{2}\left\{(b+c)\cos\theta\pm\sqrt{(b+c)^2\cos^2\theta-4bc}\right\}.$$

And $\because Ap=qp-Aq=R-r$, by giving successive values to θ, and taking the corresponding values of r, the curve, which is the locus of P, will be traced. If BD be the position of the rod when Q has described a $\frac{1}{2}\odot$, $PD=BQ$. Hence the curve is an oval, whose axis $PD=a$.

(15.) The equation to the spiral of Archimedes is $r=a\theta$; trace the curve, and show that the origin is a point of contrary flexure.

Let $\theta=0$, $\therefore r=0$,

$$\theta=45, \quad r=a\cdot\frac{3\cdot1416}{4}=a\,(\cdot7854),$$

$$\theta=\frac{\pi}{2}, \quad r=a\cdot\frac{3\cdot1416}{2}=a\,(1\cdot5708),$$

$$\theta=\pi, \quad r=a\,(3\cdot1416),$$

$$\theta=\frac{3\pi}{2}, \quad r=a\,(4\cdot7124),$$

$$\theta=2\pi, \quad r=a\,(6\cdot2832),$$

$$\theta=\infty, \quad r=\infty.$$

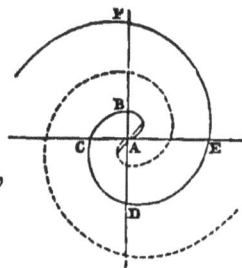

Take the angles, and draw the corresponding lines for the values of r, and the curve may be traced.

Put $-\theta$ for θ, and the values of r, being negative, must be measured in a directly contrary direction.

Now $\theta=\dfrac{r}{a}$, $\quad\therefore, \dfrac{d\theta}{dr}=\dfrac{1}{a}$. $\quad$ But $\dfrac{d\theta}{dr}=\dfrac{p}{r\sqrt{r^2-p^2}}$,

$$\therefore \frac{p}{r\sqrt{r^2-p^2}}=\frac{1}{a}, \qquad \frac{r^4-r^2p^2}{p^2}=a^2, \qquad \frac{r^4}{p^2}=a^2+r^2,$$

$$p^2=\frac{r^4}{a^2+r^2}, \qquad\qquad p=\frac{r^2}{\sqrt{a^2+r^2}}.$$

$$\therefore \frac{dp}{dr}=\frac{2r\sqrt{a^2+r^2}-r^2\cdot\dfrac{2r}{2\sqrt{a^2+r^2}}}{a^2+r^2}=\frac{2r\,(a^2+r^2)-r^3}{(a^2+r^2)^{\frac{3}{2}}}$$

$$=\frac{2a^2r+r^3}{(a^2+r^2)^{\frac{3}{2}}}=r\frac{2a^2+r^2}{(a^2+r^2)^{\frac{3}{2}}}=0, \quad \text{when } r=0 \text{ or } \theta=0;$$

and $\dfrac{dp}{dr}$ changes sign immediately before and after the origin.

∴ the origin is a point of contrary flexure.

In the figure, if r commences its revolution above the axis of x in the first quadrant, the branch of the spiral $ABCDEF$ will be generated. If negative values be given to θ, and r be measured in a directly opposite direction, the branch represented by the dotted line will be traced ; and we shall have the double spiral. If r commences its revolu-. tion upwards in the second quadrant, two branches will be generated, similar to the others, but turned in a contrary direction, and intersecting them in the horizontal and vertical axes.

This spiral was invented by Conon : but Archimedes discovered its principal properties.

If a fly were to move uniformly from the nave of a wheel along one of the spokes whilst the wheel revolved uniformly about a fixed axis, the fly would describe this spiral.

Teeth of this form are applied in the construction of engines in which uniform motion in a given direction is required.

(16.) Two points start from the opposite extremities of the diameter of a circle, and move with uniform velocity in the same direction round the circumference, their velocities are in the ratio of 2 : 1. Determine the locus of the bisection of the chords which join the positions of the two points, and find the polar subtangent of the curve.

Let the diameter $AB = 2a$, and A be the position of the point which moves with a velocity equal to double that of the point at B. Now when this latter point has made

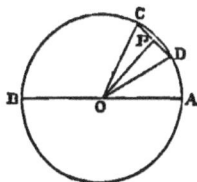

half a revolution, the former will have made a complete revolution, and consequently the two points will coincide at A. Again, the motions continuing, if we take any arc AC, and bisect it in D, C will be a position of the point which started from A, and D the corresponding position of the point which started from B. Draw the chord CD, bisect it in P, and join OP, OC, OD.

Let O be the pole, OP the radius vector $=r$, $\angle AOP=\theta$, then $POD=\dfrac{\theta}{3}$, $\dfrac{OP}{OD}=\cos POD$, or $\dfrac{r}{a}=\cos\dfrac{1}{3}\theta$, the equation to the locus of P.

To find the polar subtangent,

$$\cos\frac{1}{3}\theta=\frac{r}{a}, \qquad -\sin\frac{1}{3}\theta\cdot\frac{d\theta}{dr}=\frac{1}{a},$$

$$\frac{d\theta}{dr}=-\frac{1}{a\sin\frac{1}{3}\theta}=-\frac{1}{a\sqrt{1-\cos^2\frac{1}{3}\theta}}=-\frac{1}{a\sqrt{1-\frac{r^2}{a^2}}},$$

$$\therefore r^2\frac{d\theta}{dr}=-\frac{r^2}{\sqrt{a^2-r^2}}=\text{the polar subtangent.}$$

To trace the curve, $\qquad r=a\cos\dfrac{1}{3}\theta.$

Put $\theta=0$, then $\cos 0=1$, $\qquad r=a,$

$\theta=45$, $\qquad \cos 15=\dfrac{\sqrt{3}+1}{2\sqrt{2}}$, $\qquad r=a\dfrac{\sqrt{3}+1}{2\sqrt{2}}$,

$\theta=90$, $\qquad \cos 30=\dfrac{\sqrt{3}}{2}$, $\qquad r=a\dfrac{\sqrt{3}}{2}$,

$\theta=135$, $\qquad \cos 45=\dfrac{1}{\sqrt{2}}$, $\qquad r=\dfrac{a}{\sqrt{2}}$,

$\theta=180$, $\qquad \cos 60=\dfrac{1}{2}$, $\qquad r=\dfrac{a}{2}$,

Let $\theta=225$, then $\cos 75=\dfrac{\sqrt{3}-1}{2\sqrt{2}}$, $r=a\dfrac{\sqrt{3}-1}{2\sqrt{2}}$,

$\theta=270$, $\cos 90=0$, $r=0$,

$\theta=315$, $\cos 105=-\dfrac{\sqrt{3}-1}{2\sqrt{2}}$, $r=-a\dfrac{\sqrt{3}-1}{2\sqrt{2}}$,

$\theta=360$, $\cos 120=-\dfrac{1}{2}$, $r=-\dfrac{a}{2}$,

$\theta=405$, $\cos 135=-\dfrac{1}{\sqrt{2}}$, $r=-\dfrac{a}{\sqrt{2}}$,

$\theta=450$, $\cos 150=-\dfrac{\sqrt{3}}{2}$, $r=-a\dfrac{\sqrt{3}}{2}$,

$\theta=495$, $\cos 165=-\dfrac{\sqrt{3}+1}{2\sqrt{2}}$, $r=-\dfrac{\sqrt{3}+1}{2\sqrt{2}}$,

$\theta=540$, $\cos 180=-1$, $r=-a$.

The negative values of r, which are measured in an opposite direction, are distinguished in the figure by dotted lines.

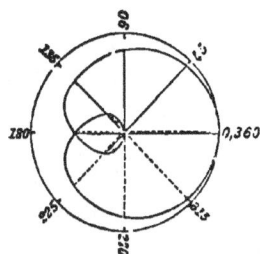

By giving negative values to θ the same curve would be produced, but turned in a contrary direction.

(17.) If $a^2y=3bx^2-x^3$; show that there is a point of contrary flexure when $x=b$, and $y=\dfrac{2b^3}{a^2}$.

(18.) If $y=2a\sqrt{\dfrac{2a-x}{x}}$, show that there are two points of inflexion when $x=\dfrac{3a}{2}$, $y=\pm\dfrac{2a}{\sqrt{3}}$.

(19.) If $ax^{\frac{1}{2}}-(x-a)y^{\frac{1}{2}}=0$ be the equation to a curve; show that there is a point of contrary flexure when $x=-2a$.

(20.) If $y = ax + bx^2 - cx^3$; show that there is a point of inflexion when $x = \dfrac{b}{3c}$, and $y = \dfrac{b}{27c^2}(9ac + 2b^2)$.

(21.) If $y = c + (x-a)^2(x-b)^{\frac{1}{3}}$; show that there is a double point when $x = a$, and $y = c$.

(22.) If $y = \dfrac{x^2}{a^3}(a^2 - x^2)$; show that there are points of inflexion when $x = \pm\dfrac{a}{\sqrt{6}}$, $y = \dfrac{5a}{6^2}$.

(23.) If $y = \dfrac{x^2}{a}\cdot\dfrac{x+a}{x-a}$, $y = x\left(\dfrac{x+a}{x-a}\right)^2$, $y = \dfrac{a^2(x^2 - a^2)^{\frac{1}{4}}}{(x-b)(x-c)}$ be three equations having no mutual relation, and x becomes infinitely great in each; prove that in (1) $y = \infty$, and $\dfrac{dy}{dx} = \infty$, in (2) $y = \infty$, and $\dfrac{dy}{dx} = 1$, and in (3) $y = 0$, and $\dfrac{dy}{dx} = 0$.

(24.) If $y^2(x^2 - a^2) = x^4$; show that the equations to the asymptotes are $y = +x$, $y = -x$, and that the curve lies above the asymptote: also show that the curve has two branches touching the axis of x at the origin, both being in a plane perpendicular to the plane of the paper, between two asymptotes which cut the axis of x at right-angles when $x = +a$, $x = -a$; show that beyond these asymptotes the curve is in the plane of reference, and approaches nearest to the axis of x when $x = a\sqrt{2}$, again receding towards the asymptotes whose equations are $y = \pm x$, and intersecting them at ∞ in a point of inflexion.

(25.) If $y^3 + x^3 - 2ax^2 = 0$; show that the equation to the asymptote is $y = -x + \dfrac{2a}{3}$, that at the origin there is a cusp of the first species, the two branches being above the axis of x and concave to it, that the curve cuts the axis of x at

M

right-angles at a point denoted by $x=2a$, where there is a point of inflexion, beyond which it approaches the asymptote whose equation is $y=-x+\dfrac{2a}{3}$; show also that there is a maximum ordinate whose length is $\dfrac{2a}{3}\cdot\sqrt[3]{4}$, when $x=\dfrac{4a}{3}$.

(26.) If $r=\dfrac{a}{\sqrt{\theta}}$; show that $p=\dfrac{2a^4 r}{\sqrt{4a^4+r^4}}$, and that there is a point of inflection when $r=a\sqrt{2}$, the curve being concave towards the pole when r is less than $a\sqrt{2}$, and convex towards it when r is greater than $a\sqrt{2}$.

(27.) $y=a+a^{\frac{1}{3}}(x-a)^{\frac{2}{3}}$; determine the nature and position of the cusp.

(28.) $y^2=\dfrac{x^4}{a^2-x^2}$ being the equation to a curve referred to rectangular co-ordinates ; show that the equation between polar co-ordinates is $r=a\tan\theta$, and that the equation between the radius vector and the perpendicular from the pole upon the tangent is $p=\dfrac{ar^2}{\sqrt{a^4+3\,a^2r^2+r^4}}$; show also how the branches of the curve are situated with regard to the plane of reference.

(29.) If $\theta=\dfrac{a}{r-a}$; show that a line drawn parallel to the prime radius or axis, at the distance a above it, is an asymptote to the curve, that, when θ is $+$, the curve has an interior asymptotic circle, and when θ is $-$, it has an exterior asymptotic circle. Trace the curve, and show that the rectilinear asymptote is a tangent to the asymptotic circle.

(30.) The equation to the *Cardioid* is $r=a(1+\cos\theta)$; trace the curve.

(31.) If $r = a \dfrac{\theta + \sin \theta}{\theta - \sin \theta}$; trace the curve, and show that there is an asymptotic circle, radius$=a$, and that the curve, coming from infinity, continually approaches the convex circumference of the asymptotic circle on one side of the diameter, and the concave circumference on the other side of the diameter.

(32.) The equation to a curve being $\dfrac{y}{x} = \sqrt{\dfrac{x^2 + a^2}{x^2 - a^2}}$, show that it has asymptotes, at right-angles to the axis of x, at points denoted by $x = +a$, $x = -a$, and other asymptotes cutting the axis of x at 45°, and 135°, respectively; that there are minimum ordinates when $x = \pm a \sqrt{\sqrt{2} + 1}$. Determine the value of these ordinates, and show the position and direction of the branches of this curve.

(33.) $y = a \pm (ax - a^2)^{\frac{1}{3}}$; determine the nature and position of the singular point.

(34.) $x^2 y^2 + a^2 y^2 - a^4 = 0$ is the equation to a curve; show that its asymptote coincides with the axis of x, and that there are points of inflexion above that axis at distances equal to $+ a \sqrt{\dfrac{2}{3}}$, and $- a \sqrt{\dfrac{2}{3}}$ from it, and at distances equal $+ \dfrac{a}{\sqrt{2}}$ and $- \dfrac{a}{\sqrt{2}}$ from the origin of co-ordinates.

(35.) If $x^3 - y^3 = a^3$; show that the curve cuts the axis of x at right-angles, at the distance a from the origin, that at each of these points there is an inflexion, the part of the curve between them being concave to the axis, the part to the left of the origin being convex, and the part to the right of the point denoted by $x = a$, concave.

(36.) If $(x-a)^5=(y-x)^2$; show that the common tangent to the two branches of the curve is inclined to the axis of x at an angle of 45°, that the curve cannot extend to the left of the point denoted by $x=a$, and that, at the distance a above that point, there is a cusp of the first species.

(37.) If $y=\sqrt[3]{\dfrac{c^4-ax^3}{b}}$ be the equation to a curve; show that there is a point of inflexion at the distance $\dfrac{c^{\frac{4}{3}}}{b^{\frac{1}{3}}}$ above the origin, and another in the axis of x, at the dis-tance $\dfrac{c^{\frac{4}{3}}}{a^{\frac{1}{3}}}$ from the origin.

(38.) $y=c\sin\dfrac{x}{a}$ is the equation to the curve of sines; show that, at all the intersections of this curve with the axis of x, there are points of contrary flexure.

(39.) $y^2=a^2+x\sqrt{2a^2-x^2}$ being the equation to a curve; show that its branches intersect the axis of x at angles $=\tan^{-1}\pm\dfrac{1}{\sqrt{2}}$ and $\tan^{-1}\pm\sqrt{2}$, that there are four double points in the axes of co-ordinates, at the distance a from the origin, and that the branches form two intersecting ovals.

(40.) If $r^2=a^2\sin2\theta$; show that there is an oval in each of the first and third quadrants, and that no curve exists in either the second or fourth quadrants.

(41.) If the equation to a curve be $x^2+y^2-2\sqrt{axy}=0$; show that the axes are tangents, that $p=0$ and ∞, and that the origin is a double point.

(42.) If $\tan3\theta=-\dfrac{y}{x}$, and $\tan\theta=\dfrac{y}{a-x}$ define a curve;

show that it has a maximum ordinate at the point denoted

by $x=a\left(1-\dfrac{\sqrt{3}}{2}\right)$, and trace the curve.

(43.) Trace the curve, whose equation is $2\,ay^3+3\,a^2y^2 +2\,a^2x^2=a^4+x^4$, and determine the different angles at which it cuts the axis of x.

(44.) Transform the equation $(a-x)\,y^2=x^3$ from rectangular to polar co-ordinates, and trace the curve.

(45.) Trace the curve, whose equation is $y^3-by^2-ax^2 =0$, and determine whether it has a point of contrary flexure.

(46.) Prove that, in the logarithmic spiral, the equation to which is $r=ae^{m\theta}$, the tangent constantly makes the same angle with the radius vector.

(47.) Trace the curve, whose equation is $\dfrac{y^2}{x^2}=\dfrac{a-2x}{a-x}$, and ascertain the angles at which it cuts the axis of x.

(48.) If the hour and minute hands of a watch were of equal length, and an elastic thread, so extensible as not to impede their motions, were attached to the extremity of each index, the thread representing a straight line of variable length, from 0 to the diameter of the dial-plate; determine the polar equation to the curve which would be described by the middle point of the thread, and trace that curve.

(49.) If perpendiculars be drawn to the diameter of a circle, and from each of them a part be taken, measured from the diameter, equal to half the sine of twice the arc which it cuts off, the arc being measured from the same extremity of the diameter; show that the equation to the curve passing through the points thus determined is a lemniscata, whose equation is $y = \frac{x}{a}\sqrt{a^2 - x^2}$, and trace the curve.

(50.) If $y = \frac{x(x-c)^{\frac{1}{2}}}{a^{\frac{3}{2}}} + b$; there is an isolated point, determine its position, and exhibit the form of the curve.

(51.) In $x^2 \log x - x^2 y + y = 0$, show that the origin is a point d'arrêt; and in $y + ye^{-x} - x = 0$ a point saillant, the branch corresponding to the negative values of x starting at an angle whose tangent is 225°.

(52.) Transform $(x^2 + y^2)^3 = a^2 y^4$ to an equation between polar co-ordinates, show that the pole is a quadruple point, and exhibit the form of the curve.

(53.) Show that the curve, the equation to which is $ay^2 = (x-a)^2(x-b)$, has a singular point when $x = a$, a conjugate point if b is greater than a, and a double point if a is greater than b.

(54.) ACB is a semicircle whose diameter is AB; draw an ordinate NC and a chord AC, then NP being taken in the ordinate, always equal to the difference between the chord and the corresponding abscissa, show that the locus of P is a parabola, and that there is a maximum ordinate when the abscissa and corresponding ordinate are equal.

(55.) Show that the curve, whose equation is $y = \frac{a^2 x}{ab + x^2}$, has three points of inflexion; and that, when $x = \sqrt{ab}$, the tangent is parallel to the axis of x.

(56.) If $r = a\theta^n$; show that there are points of contrary flexure when $r = 0$, and $r = a(-n^2 - n)^{\frac{n}{2}}$; and that this equation comprehends those of the spiral of Archimedes, the lituus, the hyperbolic or reciprocal spiral, and an infinite number of spirals.

(57.) Show how the trisectrix, the equation to which is $r = a\,(2\cos\theta - 1)$, may be used to trisect an arc or angle; and explain the difference between the generation of this curve and that of the cardioid.

(58.) Prove that the angle at which the logarithmic or equiangular spiral, whose equation is $r = a^\theta$, cuts the radius, is constant, and that the radii which include equal angles are proportional.

(59.) If $x = a\,(\theta - e\sin\theta)$, and $y = a\,(1 - e\cos\theta)$ define the trochoid; show that, at a point of contrary flexure,

$$y = \frac{a^2 - b^2}{a}.$$

(60.) A circle, which continues constantly in the same plane, rolls, like a carriage wheel, along a fixed horizontal line; the curve described by a point in the circumference is the cycloid. Find the equations $\dfrac{dx}{dy} = \left(\dfrac{y}{2a - y}\right)^{\frac12}$, and $\dfrac{dy}{dx} = \left(\dfrac{2a - x}{x}\right)^{\frac12}$.

(61.) Ascertain the loci of the transcendental equations

(1) $\quad y = x^2 + \cos x \sqrt{-1}$,

(2) $\quad y = x^2 \pm \sqrt{1 - a\sec^2 x}$.

(62.) Show that, in curves referred to polar co-ordinates, s being the length of the spiral, $\dfrac{ds}{d\theta} = \dfrac{r^2}{p}$. Investigate the equation between r and θ when $p^2 = \dfrac{r^{n+2}}{r^n + a^n}$, and between p and r when $r = a\sin n\theta$.

(63.) If a, and b, be two conjugate diameters of an ellipse,

φ the angle they make with each other, and $\dfrac{\cos^2\theta}{a^2}+\dfrac{\sin^2\theta}{b^2}=\dfrac{1}{r^2}$ the polar equation to the ellipse referred to the centre; prove that $a_i^2+b_i^2=a^2+b^2$, and $a_ib_i=ab\,\operatorname{cosec}\phi$.

(64.) Trace the curve, whose equation is $ay^2=x^3-bx^2$, and determine the number and nature of its singular points.

(65.) Let BAC be a parabola, A the vertex, and BC the latus rectum; in BC take M and N equidistant from B and C, draw MD and NE perpendicular to BC, to meet the curve in D and E, draw CD cutting NE in P. Determine the equation to the locus of P, and trace the curve.

(66.) A straight line DAE, at right-angles to the diameter ACB of a circle, moves, parallel to DAE, along the diameter, whilst a line which at first lies on the radius CA, revolves with a uniform angular motion about C, intersecting the other moving line in P; show that the equation to the curve traced out by P is $y=(a-x)\cdot\tan\dfrac{\pi x}{2a}$; that the curve, which is the quadratrix of Dinostratus, has an infinite number of infinite branches intersecting the axis of x, and that the moving parallel is an asymptote to two infinite branches. Show also that, if this curve could be geometrically described, the ratio of the diameter of a circle to its circumference would be determined.

(67.) A globe, whose radius is $a-b$, vibrates in a hollow hemisphere, whose radius is a, in such a manner that a great circle of the globe coincides with a great circle of the hemisphere; determine the curve traced out by the highest point on the globe in one revolution, and exhibit the polar equation.

CHAPTER XV.

CURVATURE OF CURVED LINES. RADIUS OF CURVATURE.
EVOLUTES.

Rectangular Co-ordinates.

If the equation to the osculating circle, or circle of curva-ture, be $R^2 = (x-\alpha)^2 + (y-\beta)^2$, and if p be put for $\dfrac{dy}{dx}$, and q for $\dfrac{d^2y}{dx^2}$, R being considered *positive* when the curve is concave to the axis of x, and *negative* when the curve is convex ; then

$$R = -\frac{(1+p^2)^{\frac{3}{2}}}{q}, \qquad y-\beta = -\frac{1+p^2}{q}, \qquad x-\alpha = \frac{1+p^2}{q}\cdot p.$$

α and β, being the co-ordinates of the centre of the radius of curvature, are the co-ordinates of the evolute of the curve.

If $u=0$ be the equation to the curve,

$$\frac{1}{R} = \frac{\left(\dfrac{du}{dy}\right)^2 \dfrac{d^2u}{dx^2} - 2\dfrac{du}{dx}\dfrac{du}{dy}\dfrac{d^2u}{dxdy} + \left(\dfrac{du}{dx}\right)^2 \dfrac{d^2u}{dy^2}}{\left\{\left(\dfrac{du}{dx}\right)^2 + \left(\dfrac{du}{dy}\right)^2\right\}^{\frac{3}{2}}},$$

The middle term of the numerator in this expression vanishes when the value of u is the sum of two parts, one involving x and the other y.

The distance from a point in the curve to the intersection of two consecutive normals is the radius of curvature at that point.

The normal to the curve is the tangent to the evolute.

Polar Co-ordinates.

If R be the radius of curvature as before, r the radius vector, θ the angle traced out by r, and p the perpendicular upon the tangent,

$$R = r\frac{dr}{dp} = \frac{\left(r^2 + \dfrac{dr^2}{d\theta^2}\right)^{\frac{3}{2}}}{r^2 + 2\dfrac{dr^2}{d\theta^2} - r\dfrac{d^2r}{d\theta^2}}.$$

$$\text{The semi-chord} = p\frac{dr}{dp} = \frac{r\left(\dfrac{dr^2}{d\theta^2} + r^2\right)}{r^2 + 2\dfrac{dr^2}{d\theta^2} - r\dfrac{d^2r}{d\theta^2}}.$$

To find the equation to the evolute to a spiral; r and p being taken as co-ordinates of the involute, $r_{,}$ and $p_{,}$ as corresponding co-ordinates of the evolute, we must eliminate R, r and p from the four equations

$$p = f(r), \quad p_{,} = (r^2 - p^2)^{\frac{1}{2}}, \quad R = r\frac{dr}{dp}, \quad r_{,}^2 = r^2 + R^2 - 2Rp.$$

Ex. (1.) To determine the radius of curvature at any point in the common parabola.

$$y^2 = 4mx, \quad \text{the equation to the curve,}$$

$$2y\frac{dy}{dx} = 4m, \qquad \therefore p = \frac{dy}{dx} = \frac{2m}{y},$$

$$q = \frac{d^2y}{dx^2} = -\frac{2m}{y^2} \cdot \frac{dy}{dx} = -\frac{2m}{y^2} \cdot \frac{2m}{y} = -\frac{4m^2}{y^3},$$

$$1 + p^2 = 1 + \frac{4m^2}{y^2} = \frac{4m^2 + y^2}{y^2} = \frac{4m^2 + 4mx}{y^2},$$

$$\therefore R = -\frac{(1 + p^2)^{\frac{3}{2}}}{q} = \frac{\{4m(m+x)\}^{\frac{3}{2}}}{y^3} \cdot \frac{y^3}{4m^2} = \frac{2(m+x)^{\frac{3}{2}}}{m^{\frac{1}{2}}}.$$

Since this expression for the radius of curvature diminishes as x diminishes, R is least when $x = 0$, and then $R = 2m$

= half the latus rectum ; hence in the parabola the point of greatest curvature is the vertex.

(2.) The equation to the rectangular hyperbola, referred to its asymptotes is $xy=m^2$; find the radius of curvature.

$$x\frac{dy}{dx}+y=0, \qquad x\frac{dy}{dx}=-\frac{m^2}{x}, \qquad \therefore p=-\frac{m^2}{x^2},$$

$$q=\frac{d^2y}{dx^2}=\frac{2m^2x}{x^4}=\frac{2m^2}{x^3},$$

$$1+p^2=1+\frac{m^4}{x^4}=\frac{x^4+m^4}{x^4}=\frac{x^4+x^2y^2}{x^4}=\frac{x^2+y^2}{x^2},$$

$$\therefore R=-\frac{(1+p^2)^{\frac{3}{2}}}{q}=-\frac{(x^2+y^2)^{\frac{3}{2}}}{x^3}\cdot\frac{x^3}{2m^2}=-\frac{(x^2+y^2)^{\frac{3}{2}}}{2m^2}.$$

(3.) If the equation to a circle be $x^2-a(x-y)+y^2=0$; find the radius of curvature.

$$y^2+ay=ax-x^2, \qquad (2y+a)\frac{dy}{dx}=a-2x,$$

$$\therefore p=\frac{a-2x}{a+2y}, \qquad p^2=\frac{(a-2x)^2}{(a+2y)^2},$$

$$1+p^2=1+\frac{(a-2x)^2}{(a+2y)^2}=\frac{(a+2y)^2+(a-2x)^2}{(a+2y)^2},$$

$$q=\frac{-2(a+2y)-2(a-2x)\cdot p}{(a+2y)^2}=\frac{-2(a+2y)-2(a-2x)\cdot\frac{a-2x}{a+2y}}{(a+2y)^2}$$

$$\therefore -q=\frac{2\{(a+2y)^2+(a-2x)^2\}}{(a+2y)^3}.$$

Now $R=-\frac{(1+p^2)^{\frac{3}{2}}}{q}=\frac{\{(a+2y)^2+(a-2x)^2\}^{\frac{3}{2}}}{2\{(a+2y)^2+(a-2x)^2\}}$

$$=\frac{\{(a+2y)^2+(a-2x)^2\}^{\frac{1}{2}}}{2}=\frac{(2a^2)^{\frac{1}{2}}}{2}=\frac{a}{2^{\frac{1}{2}}}.$$

(4.) Find the radius of curvature to the hyperbola, and determine the equation to its evolute.

$$y^2 = \frac{b^2}{a^2}(x^2 - a^2), \quad \text{the equation to the curve,}$$

$$y\frac{dy}{dx} = \frac{b^2 x}{a^2}, \qquad \therefore p = \frac{dy}{dx} = \frac{b^2 x}{a^2 y}, \qquad 1 + p^2 = 1 + \frac{b^4 x^2}{a^4 y^2},$$

$$\therefore 1 + p^2 = 1 + \frac{b^4 x^2}{a^2 b^2 x^2 - a^4 b^2} = 1 + \frac{b^2 x^2}{a^2 x^2 - a^4} = \frac{a^2 x^2 + b^2 x^2 - a^4}{a^2 x^2 - a^4},$$

$$(1 + p^2)^{\frac{3}{2}} = \frac{(a^2 x^2 + b^2 x^2 - a^4)^{\frac{3}{2}}}{(a^2 x^2 - a^4)^{\frac{3}{2}}},$$

$$q = \frac{d^2 y}{dx^2} = \frac{a^2 y b^2 - b^2 x \cdot a^2 p}{a^4 y^2} = \frac{b^2 y - b^2 x \cdot \dfrac{b^2 x}{a^2 y}}{a^2 y^2} = \frac{a^2 b^2 y^2 - b^4 x^2}{a^4 y^3}$$

$$= \frac{a^2 b^2 \cdot \dfrac{b^2}{a^2}(x^2 - a^2) - b^4 x^2}{a^4 \cdot \dfrac{b^3}{a^3}(x^2 - a^2)^{\frac{3}{2}}} = \frac{b^4 x^2 - a^2 b^4 - b^4 x^2}{ab^3 (x^2 - a^2)^{\frac{3}{2}}} = -\frac{ab}{(x^2 - a^2)^{\frac{3}{2}}}.$$

Hence $R = -\dfrac{(1 + p^2)^{\frac{3}{2}}}{q} = \dfrac{(a^2 x^2 + b^2 x^2 - a^4)^{\frac{3}{2}}}{a^3 (x^2 - a^2)^{\frac{3}{2}}} \cdot \dfrac{(x^2 - a^2)^{\frac{3}{2}}}{ab}$

$$= \frac{(a^2 x^2 + b^2 x^2 - a^4)^{\frac{3}{2}}}{a^4 b} = \frac{\{a^2 x^2 + a^2 (e^2 - 1) x^2 - a^4\}^{\frac{3}{2}}}{a^4 b}$$

$$= \frac{(a^2 e^2 x^2 - a^4)^{\frac{3}{2}}}{a^4 b} = \frac{\{a^2 (e^2 x^2 - a^2)\}^{\frac{3}{2}}}{a^4 b} = \frac{a^3 (e^2 x^2 - a^2)^{\frac{3}{2}}}{a^4 b}$$

$$= \frac{(e^2 x^2 - a^2)^{\frac{3}{2}}}{ab} = \text{radius of curvature.}$$

To find the equation to the evolute,

$$y - \beta = -\frac{1 + p^2}{q} = \frac{a^2 x^2 + b^2 x^2 - a^4}{a^2 (x^2 - a^2)} \cdot \frac{(x^2 - a^2)^{\frac{3}{2}}}{ab}$$

$$= \frac{\{a^2 x^2 + a^2 (e^2 - 1) x^2 - a^4\}(x^2 - a^2)^{\frac{1}{2}}}{a^3 b}$$

$$= \frac{(a^2 e^2 x^2 - a^4)(x^2 - a^2)^{\frac{1}{2}}}{a^3 b} = \frac{(e^2 x^2 - a^2)(x^2 - a^2)^{\frac{1}{2}}}{ab}$$

$$= \frac{(e^2 x^2 - a^2) \cdot \dfrac{ay}{b}}{ab} = \frac{y(e^2 x^2 - a^2)}{b^2}.$$

$$\therefore \beta = -\frac{y(e^2x^2-a^2)}{b^2} + y = -y\left\{\frac{e^2x^2-a^2}{b^2}-1\right\}$$

$$= -y\left\{\frac{e^2x^2-a^2-b^2}{b^2}\right\} = -\frac{y}{b^2}\{e^2x^2-a^2-a^2(e^2-1)\}$$

$$= -\frac{y}{b^2}\{e^2x^2-e^2a^2\} = -\frac{ye^2}{b^2}(x^2-a^2) = -\frac{ye^2}{b^2}\cdot\frac{a^2y^2}{b^2}$$

$$= -\frac{y^3}{b^3}\cdot\frac{e^2a^2}{b}.$$

$$\therefore \frac{y^3}{b^3} = -\frac{b\beta}{(ae)^2}, \qquad \frac{y^2}{b^2} = \cdot\frac{(b\beta)^{\frac{2}{3}}}{(ae)^{\frac{4}{3}}}.$$

$$x-a = -p(y-\beta) = -\frac{b^2x}{a^2y}\cdot\frac{y(e^2x^2-a^2)}{b^2} = -\frac{x}{a^2}(e^2x^2-a^2)$$

$$= x - \frac{e^2x^3}{a^2}, \quad \therefore a = \frac{x^3}{a^2}\cdot e^2, \quad \frac{x^3}{a^3} = \frac{a}{ae^2}, \quad \frac{x^2}{a^2} = \frac{a^{\frac{2}{3}}}{a^{\frac{2}{3}}e^{\frac{4}{3}}}.$$

But $\frac{y^2}{b^2} - \frac{x^2}{a^2} = -1,$ $\qquad \therefore \frac{(b\beta)^{\frac{2}{3}}}{(ae)^{\frac{4}{3}}} - \frac{a^{\frac{2}{3}}}{a^{\frac{2}{3}}e^{\frac{4}{3}}} = -1.$

$$\therefore (b\beta)^{\frac{2}{3}} - (a\alpha)^{\frac{2}{3}} = -(ae)^{\frac{4}{3}} = -(a^2e^2)^{\frac{2}{3}} = -(a^2+b^2)^{\frac{2}{3}}.$$

$$\therefore (a\alpha)^{\frac{2}{3}} - (b\beta)^{\frac{2}{3}} = (a^2+b^2)^{\frac{2}{3}} \quad \text{the equation to the evolute.}$$

(5.) Show that, in the catenary, the radius is equal but opposite to the normal.

$$y = \frac{a}{2}(e^{\frac{x}{a}} + e^{-\frac{x}{a}}), \text{ the equation to the curve,}$$

$$p = \frac{dy}{dx} = \frac{a}{2}\left(\frac{e^{\frac{x}{a}}}{a} - \frac{e^{-\frac{x}{a}}}{a}\right) = \frac{e^{\frac{x}{a}} - e^{-\frac{x}{a}}}{2},$$

$$1+p^2 = 1 + \frac{e^{\frac{2x}{a}}-2+e^{-\frac{2x}{a}}}{4} = \frac{e^{\frac{2x}{a}}+2+e^{-\frac{2x}{a}}}{4} = \left(\frac{e^{\frac{x}{a}}+e^{-\frac{x}{a}}}{2}\right)^2 = \frac{y^2}{a^2}.$$

$$q = \frac{d^2y}{dx^2} = \frac{2a^2y}{a^4} = \frac{2y}{a^2}.$$

$$\therefore R = -\frac{(1+p^2)^{\frac{3}{2}}}{q} = \frac{y^3}{a^3}\cdot\left(-\frac{a^2}{2y}\right) = -\frac{y^2}{a}.$$

But the normal $N = y \sqrt{1 + \dfrac{dy^2}{dx^2}} = y \sqrt{1 + p^2} = y \cdot \dfrac{y}{a} = + \dfrac{y^2}{a}.$

Hence the radius of curvature is equal but opposite to the normal.

(6.) Determine the radius of curvature and the evolute of the cycloid.

Let $AN = x$, $NP = y$, $CD = 2a$.

$\dfrac{y}{a} = \text{versin } \dfrac{x + \sqrt{2ay - y^2}}{a}$,   the equation to the curve.

$p = \dfrac{dy}{dx} = \dfrac{\sqrt{2a - y}}{\sqrt{y}},$ $p^2 = \dfrac{2a}{y} - 1,$

$$1 + p^2 = \dfrac{2a}{y},$$

$2p \dfrac{d^2 y}{dx^2} = -\dfrac{2a}{y^2} \cdot \dfrac{dy}{dx},$ $\therefore q = \dfrac{d^2 y}{dx^2} = -\dfrac{a}{y^2}.$

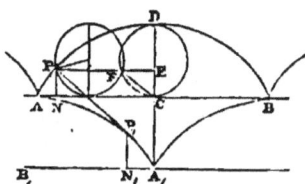

Hence $R = -\dfrac{(1 + p^2)^{\frac{3}{2}}}{q} = \left(\dfrac{2a}{y}\right)^{\frac{3}{2}} \cdot \dfrac{y^2}{a} = 2\sqrt{2ay}.$

Now $CF^2 = CE^2 + EF^2 = CE^2 + CE \cdot ED = y^2 + y(2a - y),$

$\therefore CF = \sqrt{2ay},$ $\therefore R = 2CF = $ radius of curvature.

To find the equation to the evolute,

$y - \beta = -\dfrac{1 + p^2}{q} = \dfrac{2a}{y} \cdot \dfrac{y^2}{a} = 2y,$ $\therefore \beta = -y,$

$x - \alpha = -p(y - \beta) = -\dfrac{\sqrt{2a - y}}{\sqrt{y}} \cdot 2y = -2\sqrt{2ay - y^2},$

$$\therefore \alpha = x + 2\sqrt{2ay - y^2}.$$

Substituting these values of α and β for the co-ordinates in the equation to the curve, we have

$$-\dfrac{\beta}{a} = \text{versin } \dfrac{\alpha - \sqrt{-2a\beta - \beta^2}}{a}. \qquad \ldots \ldots (1)$$

Taking $CA_{,}=CD$, and $A_{,}B_{,}$, parallel to AB, as the axis of the abscissæ, and substituting $\beta_{,}-2a$ for β, and $\pi a - a_{,}$ for a, πa being equal to AC; the origin will be transferred to $A_{,}$, and equation (1) will become

$$2 - \frac{\beta_{,}}{a} = \text{versin} \left(\pi - \frac{a_{,} + \sqrt{2 a \beta_{,} - \beta_{,}^2}}{a} \right),$$

$$\therefore \frac{\beta_{,}}{a} = \text{versin} \frac{a_{,} + \sqrt{2 a \beta_{,} - \beta_{,}^2}}{a}, \qquad \because 2 - \text{vers} A = \text{vers}(\pi - A).$$

And $\because A_{,}N_{,}=a_{,}$, and $N_{,}P_{,}=\beta_{,}$, this equation to the evolute is the equation to another cycloid originating at $A_{,}$, and whose generating circle is equal to that of the given cycloid, but moves in an opposite direction.

(7.) Show that, in the common parabola, the chord of curvature through the focus is equal to four times the focal distance; and find the length of the evolute in terms of the focal distance and the distance between the focus and vertex.

Let the focal distance $SP=r$, the perpendicular from the focus upon the tangent, $SY=p$, and $DS=2SA=2a=c$.

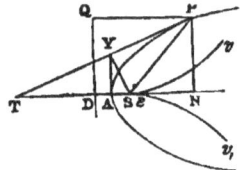

Then, by a property of the parabola, $SY^2=SP \cdot SA$,

$$\text{or} \quad p^2=\frac{cr}{2}, \qquad \therefore 2p\frac{dp}{dr}=\frac{c}{2}, \qquad \frac{dr}{dp}=\frac{4p}{c},$$

$$\text{Chord} = 2p \cdot \frac{dr}{dp}=\frac{8p^2}{c}=\frac{8}{c} \cdot \frac{cr}{2}=4r=4SP.$$

Again, $y^2=4ax$, the equation to the curve,

$$\therefore \frac{dy}{dx}=\frac{2a}{y}, \qquad \therefore p^2=\frac{4a^2}{y^2},$$

$$1+p^2=1+\frac{4a^2}{y^2}=\frac{4a^2+y^2}{y^2}=\frac{4a(a+x)}{y^2}.$$

$$q=\frac{d^2y}{dx^2}=-\frac{2a}{y^2}\cdot\frac{dy}{dx}=-\frac{2a}{y^2}\cdot\frac{2a}{y}=-\frac{4a^2}{y^3},$$

$$\therefore R=-\frac{(1+p^2)^{\frac{3}{2}}}{q}=\frac{\{4a(a+x)\}^{\frac{3}{2}}}{y^3}\cdot\frac{y^3}{4a^2}=\frac{2(a+x)^{\frac{3}{2}}}{a^{\frac{1}{2}}}=\frac{2SP^{\frac{3}{2}}}{SA^{\frac{1}{2}}}.$$

Hence,   length of evolute $s=R-c=\dfrac{2SP^{\frac{3}{2}}}{SA^{\frac{1}{2}}}-2SA$

$$=\frac{2(SP^{\frac{3}{2}}-SA^{\frac{3}{2}})}{SA^{\frac{1}{2}}}.$$

The form of the evolute, which is a semi-cubical parabola, is represented in the figure, by the lines ev, ev_1.

(8.) Find the value of the radius vector in the spiral of Archimedes, when the radius of curvature equals the chord of curvature.

$$r=a\theta, \qquad \text{the equation to the curve,}$$

$$\frac{dr}{d\theta}=a. \qquad \text{But } \frac{dr}{d\theta}=\frac{r\sqrt{r^2-p^2}}{p}.$$

$$\therefore \frac{r\sqrt{r^2-p^2}}{p}=a, \qquad \frac{r^4-r^2p^2}{p^2}=\frac{r^4}{p^2}-r^2=a^2.$$

$$r^4-p^2r^2=a^2p^2,$$

$$4r^3\cdot\frac{dr}{dp}-2pr^2-2p^2r\cdot\frac{dr}{dp}=2a^2p,$$

$$2r(2r^2-p^2)\frac{dr}{dp}=p(r^2+a^2), \qquad r\frac{dr}{dp}=\frac{p(r^2+a^2)}{2(2r^2-p^2)}.$$

But $\dfrac{r^4}{p^2}=r^2+a^2,$ $\therefore p^2=\dfrac{r^4}{r^2+a^2},$ $p=\dfrac{r^2}{(r^2+a^2)^{\frac{1}{2}}}.$

Hence $R=r\dfrac{dr}{dp}=\dfrac{\dfrac{r^2}{(r^2+a^2)^{\frac{1}{2}}}\cdot(r^2+a^2)}{2\left(2r^2-\dfrac{r^4}{r^2+a^2}\right)}=\dfrac{(r^2+a^2)^{\frac{3}{2}}}{2(r^2+2a^2)}.$

Now, chord $= 2p\dfrac{dr}{dp} = 2p\cdot\dfrac{p(r^2+a^2)}{2r(2r^2-p^2)} = \dfrac{p^2(r^2+a^2)}{r(2r^2-p^2)}$

$$= \dfrac{\dfrac{r^4}{r^2+a^2}\cdot(r^2+a^2)}{r\left(2r^2-\dfrac{r^4}{r^2+a^2}\right)} = \dfrac{2r(r^2+a^2)}{2(r^2+2a^2)}.$$

And, comparing this value of the chord with the value of the radius of curvature, already determined, it appears that radius = chord if $(r^2+a^2)^{\frac{3}{2}} = 2r(r^2+a^2)$, or

$$(r^2+a^2)^{\frac{1}{2}} = 2r, \qquad r^2+a^2 = 4r^2, \qquad 3r^2 = a^2, \qquad \therefore r = \dfrac{a}{\sqrt{3}}.$$

(9.) To find the radius of curvature in the semi-cubical parabola.

$$y^2 = \dfrac{2x^3}{3a}, \qquad \text{the equation to the curve,}$$

$$2y\cdot\dfrac{dy}{dx} = \dfrac{2x^2}{a}, \qquad \therefore p = \dfrac{dy}{dx} = \dfrac{x^2}{ay},$$

$$1+p^2 = 1+\dfrac{x^4}{a^2y^2} = \dfrac{a^2y^2+x^4}{a^2y^2} = \dfrac{2ax^3+3x^4}{3a^2y^2}.$$

$$q = \dfrac{d^2y}{dx^2} = \dfrac{2axy-ax^2\cdot p}{a^2y^2} = \dfrac{2axy-ax^2\cdot\dfrac{x^2}{ay}}{a^2y^2}$$

$$= \dfrac{2axy^2-x^4}{a^2y^3} = \dfrac{\dfrac{4x^4}{3}-x^4}{a^2y^3} = \dfrac{4x^4-3x^4}{3a^2y^3} = \dfrac{x^4}{3a^2y^3}.$$

Now $R = -\dfrac{(1+p^2)^{\frac{3}{2}}}{q} = -\left(\dfrac{2ax^3+3x^4}{3a^2y^2}\right)^{\frac{3}{2}}\cdot\dfrac{3a^2y^3}{x^4}$

$$= -\dfrac{(2a+3x)^{\frac{3}{2}}x^{\frac{9}{2}}}{3^{\frac{3}{2}}a^3y^3}\cdot\dfrac{3a^2y^3}{x^4} = -\dfrac{(2a+3x)^{\frac{3}{2}}x^{\frac{1}{2}}}{3^{\frac{1}{2}}a}.$$

(10.) Find the radius of curvature and chord of curvature in the cardioid.

$$r=a\,(1+\cos\theta), \qquad \text{the equation to the curve,}$$

$$\frac{dr}{d\theta}=-a\sin\theta, \qquad \frac{d\theta}{dr}=-\frac{1}{a\sin\theta}. \qquad \text{But} \ \ \frac{d\theta}{dr}=\frac{p}{r\sqrt{r^2-p^2}},$$

$$\therefore \ \frac{p}{r\sqrt{r^2-p^2}}=-\frac{1}{a\sin\theta}, \qquad \frac{p^2}{r^4-r^2p^2}=\frac{1}{a^2\sin^2\theta}=\frac{1}{a^2-a^2\cos^2\theta}.$$

But $\because \ a\cos\theta=r-a, \qquad a^2\cos^2\theta=r^2-2ar+a^2,$

$$\therefore \ \frac{r^4-r^2p^2}{p^2}=\frac{r^4}{p^2}-r^2=a^2-(r^2-2ar+a^2),$$

$$\therefore \ \frac{r^4}{p^2}=2ar, \qquad r^3=2ap^2, \qquad 3r^2\frac{dr}{dp}=4a\cdot p=4a\cdot\frac{r^2}{\sqrt{2ar}}.$$

Hence $\qquad R=r\cdot\dfrac{dr}{dp}=\dfrac{2\times 2ar}{3\sqrt{2ar}}=\dfrac{2}{3}\sqrt{2ar}.$

Chord $=2p\cdot\dfrac{dr}{dp}=2p\cdot\dfrac{4ap}{3r^2}=2\cdot\dfrac{4a}{3r^2}\cdot p^2=2\,\dfrac{4a}{3r^2}\cdot\dfrac{r^3}{2a}=\dfrac{4}{3}r.$

(11.) If R and $R_{\text{,}}$ respectively represent the radii of curvature of an ellipse at the extremities of two conjugate diameters; show that $\quad R^{\frac{2}{3}}+R_{\text{,}}^{\frac{2}{3}}=\sqrt[3]{\dfrac{a^4}{b^2}}+\sqrt[3]{\dfrac{b^4}{a^2}}.$

Let Pp, Qq be two diameters, then if the tangent at Q be parallel to Pp, or if the tangent at P be parallel to Qq, they will be conjugate diameters.

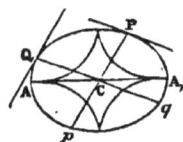

Let $CP=r, \qquad \angle PCA_{\text{,}}=\theta, \qquad CQ=r_{\text{,}}, \qquad \angle QCA=\phi.$

Then $p^2=\dfrac{a^2b^2}{a^2+b^2-r^2}\ \cdots\ (1) \qquad p_{\text{,}}^2=\dfrac{a^2b^2}{a^2+b^2-r_{\text{,}}^2}\ \cdots\ (2),$

(1) $\quad 2\log p=\log a^2b^2-\log\,(a^2+b^2-r^2),$

$$\frac{2}{p}\cdot\frac{dp}{dr}=\frac{2r}{a^2+b^2-r^2}, \qquad \frac{1}{r}\cdot\frac{dp}{dr}=\frac{p}{a^2+b^2-r^2}.$$

$$\therefore R = r\frac{dr}{dp} = \frac{a^2 + b^2 - r^2}{p} = (a^2 + b^2 - r^2)\cdot\frac{(a^2+b^2-r^2)^{\frac{1}{2}}}{ab}$$

$$= \frac{(a^2 + b^2 - r^2)^{\frac{3}{2}}}{ab}.$$

$$R_{,} = r_{,}\frac{dr_{,}}{dp_{,}} = \frac{(a^2 + b^2 - r_{,}^2)^{\frac{3}{2}}}{ab}.$$

$$\therefore R^{\frac{2}{3}} = \frac{a^2 + b^2 - r^2}{(ab)^{\frac{2}{3}}}, \qquad R_{,}^{\frac{2}{3}} = \frac{a^2 + b^2 - r_{,}^2}{(ab)^{\frac{2}{3}}}.$$

Hence $\qquad R^{\frac{2}{3}} + R_{,}^{\frac{2}{3}} = \dfrac{2(a^2 + b^2) - (r^2 + r_{,}^2)}{(ab)^{\frac{2}{3}}}.$

But since, in an ellipse, the sum of the squares of any two conjugate diameters is equal to the sum of the squares of the major and minor axes, therefore $(2a)^2 + (2b)^2 = (2r)^2 + (2r_{,})^2$,
or $a^2 + b^2 = r^2 + r_{,}^2$,

$$\therefore R^{\frac{2}{3}} + R_{,}^{\frac{2}{3}} = \frac{a^2 + b^2}{a^{\frac{2}{3}} b^{\frac{2}{3}}} = \frac{a^{\frac{4}{3}}}{b^{\frac{2}{3}}} + \frac{b^{\frac{4}{3}}}{a^{\frac{2}{3}}} = \sqrt[3]{\frac{a^4}{b^2}} + \sqrt[3]{\frac{b^4}{a^2}}.$$

The form of the evolute of an ellipse is represented in the figure.

(12.) Find the equation to the evolute of the logarithmic curve.

$$y = ae^{\frac{x}{a}}, \qquad \text{the equation to the curve,}$$

$$p = \frac{dy}{dx} = ae^{\frac{x}{a}}\cdot\frac{1}{a} = e^{\frac{x}{a}} = \frac{y}{a}, \qquad q = \frac{d^2y}{dx^2} = \frac{1}{a}\frac{dy}{dx} = \frac{1}{a}\cdot\frac{y}{a} = \frac{y}{a^2},$$

$$1 + p^2 = 1 + \frac{y^2}{a^2} = \frac{a^2 + y^2}{a^2}.$$

Now $\quad y - \beta = -\dfrac{1 + p^2}{q} = -\dfrac{a^2 + y^2}{a^2}\cdot\dfrac{a^2}{y} = -\dfrac{a^2 + y^2}{y},$

$$y^2 - \beta y = -a^2 - y^2, \qquad 2y^2 - \beta y = -a^2, \qquad y^2 - \frac{\beta}{2}y = -\frac{a^2}{2},$$

$$y^2 - \frac{\beta}{2}y + \overline{\frac{\beta}{4}}\Big|^2 = \frac{\beta^2 - 8a^2}{16}, \qquad \therefore y = \frac{\beta \pm (\beta^2 - 8a^2)^{\frac{1}{4}}}{4}.$$

But $\dfrac{d\alpha}{d\beta}=-\dfrac{dy}{dx}=-\dfrac{y}{a},$ $\qquad y=-a\dfrac{d\alpha}{d\beta},$

$\therefore -a\dfrac{d\alpha}{d\beta}=\dfrac{\beta\pm(\beta^2-8a^2)^{\frac{1}{2}}}{4}$ is the equation to the evolute.

(13.) If D be the point of intersection of the directrix and axis of the common parabola, and PN, QM be ordinates of corresponding points in the parabola and its evolute; show that $DM=3DN$.

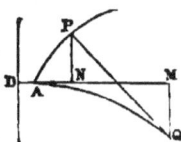

The evolute of the common parabola is the semicubical parabola.

The normal to the curve is the tangent to the evolute.

$y^2=4ax,$ the equation to the common parabola,

$\beta^2=\dfrac{4}{27a}(a-2a)^3,$ . . . . . semicubical parabola,

$y_1-y=-\dfrac{dx}{dy}(x_1-x),$ equation to the normal,

$\therefore y_1-y=-\dfrac{y}{2a}(x_1-x),$

$y_1=-\dfrac{y}{2a}x_1+\dfrac{y}{2a}x+y=-\dfrac{y}{2a}x_1+y\dfrac{x+2a}{2a}.$

Let $y=0$, then $x_1=x+2a$, the part cut off from the axis of x by the normal to the curve.

Again $2\log\beta=\log\dfrac{4}{27a}+3\log(a-2a),$ $\quad 2\dfrac{d\beta}{d\alpha}\cdot\dfrac{1}{\beta}=3\cdot\dfrac{1}{a-2a},$

$\beta\cdot\dfrac{d\alpha}{d\beta}=\dfrac{2}{3}(a-2a),$ $\quad \alpha-\beta\dfrac{d\alpha}{d\beta}=a-\dfrac{2}{3}(a-2a),$

$\therefore \alpha-\beta\dfrac{d\alpha}{d\beta}=\dfrac{a+4a}{3},$ the part cut off from the axis of x by the tangent to the evolute.

Hence $x+2a=\dfrac{a+4a}{3},$ $\qquad 3x+6a=a+4a,$

$\therefore 3x=a-2a.$

But $DN=a+x,$ $x=DN-a,$ $3x=3DN-3a,$

$DM=a+a,$ $a=DM-a,$ $a-2a=DM-3a.$

$\therefore DM-3a=3DN-3a,$ $\therefore DM=3DN.$

(14.) In an ellipse, e being the eccentricity, determine the radius of curvature in terms of the angle made by the normal with the major axis.

Normal $PG=y\sqrt{1+\dfrac{dy^2}{dx^2}},$

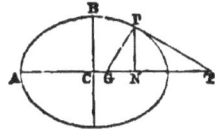

$\text{Sin } PGN=\sin\phi=\dfrac{PN}{PG}=\dfrac{y}{PG},$

$\therefore \sin\phi=\dfrac{1}{\sqrt{1+p^2}}.$

Now $y=\dfrac{b}{a}\sqrt{a^2-x^2},$ the equation to the ellipse,

$\therefore p=\dfrac{dy}{dx}=-\dfrac{b}{a}\dfrac{x}{\sqrt{a^2-x^2}},$

$1+p^2=1+\dfrac{b^2x^2}{a^2(a^2-x^2)}=\dfrac{a^4-(a^2-b^2)x^2}{a^2(a^2-x^2)}=\dfrac{a^2-e^2x^2}{a^2-x^2},$

$\therefore (1+p^2)^{\frac{3}{2}}=\dfrac{(a^2-e^2x^2)^{\frac{3}{2}}}{(a^2-x^2)^{\frac{3}{2}}},$ and $q=-\dfrac{ba}{(a^2-x^2)^{\frac{3}{2}}}.$

Hence $R=-\dfrac{(1+p^2)^{\frac{3}{2}}}{q}=\dfrac{(a^2-e^2x^2)^{\frac{3}{2}}}{ba}.$ $\ldots\ldots$ (1)

Now $\sin^2\phi=\dfrac{1}{1+p^2}=\dfrac{a^2-x^2}{a^2-e^2x^2},$ $a^2\sin^2\phi-e^2x^2\sin^2\phi=a^2-x^2,$

$(1-e^2\sin^2\phi)x^2=a^2(1-\sin^2\phi),$ $\therefore x^2=\dfrac{a^2(1-\sin^2\phi)}{1-e^2\sin^2\phi},$

$e^2x^2=\dfrac{a^2e^2(1-\sin^2\phi)}{1-e^2\sin^2\phi},$

$a^2-e^2x^2=a^2-\dfrac{a^2e^2(1-\sin^2\phi)}{1-e^2\sin^2\phi}=\dfrac{a^2(1-e^2)}{1-e^2\sin^2\phi},$

and, substituting this value of $a^2-e^2x^2$ in equation (1),

$$R=\frac{a^3(1-e^2)^{\frac{3}{2}}}{ba(1-e^2\sin^2\phi)^{\frac{3}{2}}}=\frac{a(1-e^2)^{\frac{3}{2}}}{\frac{b}{a}(1-e^2\sin^2\phi)^{\frac{3}{2}}}$$

$$=\frac{a(1-e^2)^{\frac{3}{2}}}{(1-e^2)^{\frac{1}{2}}\cdot(1-e^2\sin^2\phi)^{\frac{3}{2}}}=\frac{a(1-e^2)}{(1-e^2\sin^2\phi)^{\frac{3}{2}}}.$$

(15.) An inextensible cord AB is attached to a stone at B, and a person holding the other extremity of the cord, moves with it at right-angles to AB uniformly along the straight line AC; it is required to determine the equation to the curve described by the stone, and to find its evolute.

Let the person be supposed to move in the direction AC until he arrives at any point T, while the stone moves along the curve BP; the cord will then be in the position PT, and since up to this moment the stone has never been so near to the line AC as it now is, the line PT produced would not cut the curve BP; hence PT, or the cord in any position, is a tangent to the curve.

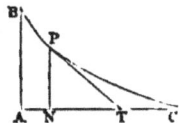

Let $AN=x$, $\quad NP=y$, $\quad AB=a$; $\quad$ then

Subtangent $NT=y\dfrac{dx}{dy}$, $\quad$ and $\quad NT^2=PT^2-NP^2$, $\quad$ or

$$y^2\left(\frac{dx}{dy}\right)^2=a^2-y^2, \quad \therefore\ y\frac{dx}{dy}=\pm\sqrt{a^2-y^2}, \quad \text{the equation}$$

required.

Hence the curve is the tractory, and AC is its directrix. ·

The equation may be readily reduced to $y\sqrt{1+\dfrac{dx^2}{dy^2}}=a$,

a form in which it is frequently given.

To determine the evolute; $\quad\left(\dfrac{dx}{dy}\right)^2=\dfrac{1}{p^2}=\dfrac{a^2}{y^2}-1,$

$$\therefore\ 1+p^2=\frac{a^2}{a^2-y^2}.$$

Again $\dfrac{dy}{dx}=\dfrac{y}{\sqrt{a^2-y^2}},$ $\therefore q=\dfrac{d^2y}{dx^2}=\dfrac{a^2y}{(a^2-y^2)^2},$

$\therefore y-\beta=-\dfrac{1+p^2}{q}=-\dfrac{a^2}{y}+y,$ $\therefore \beta=\dfrac{a^2}{y},$

$\dfrac{d\beta}{dx}=-\dfrac{a^2}{y\sqrt{a^2-y^2}},$ $x-\alpha=-(y-\beta)\dfrac{dy}{dx}=\sqrt{a^2-y^2},$

$\alpha=x-\sqrt{a^2-y^2},$ $\dfrac{d\alpha}{dx}=\dfrac{a^2}{a^2-y^2}.$

Hence $\dfrac{d\beta}{d\alpha}=\dfrac{d\beta}{dx}\cdot\dfrac{dx}{d\alpha}=-\dfrac{a^2}{y\sqrt{a^2-y^2}}\cdot\dfrac{a^2-y^2}{a^2}=-\dfrac{\sqrt{a^2-y^2}}{y}$

$=-\dfrac{\sqrt{a^2-\dfrac{a^4}{\beta^2}}}{\dfrac{a^2}{\beta}}=-\dfrac{a\sqrt{\beta^2-a^2}}{a^2}=-\dfrac{\sqrt{\beta^2-a^2}}{a},$

the equation to the evolute. Hence the evolute to the tractrix is the catenary.

(16.) The equation to a circle being $y=(a^2-x^2)^{\frac{1}{2}}$; prove that the radius of curvature equals a.

(17.) $\dfrac{x^2}{a^2}+\dfrac{y^2}{b^2}=1$ being the equation to the ellipse; show that the radius of curvature is $\dfrac{(a^2-e^2x^2)^{\frac{3}{2}}}{ab}$, where the eccentricity $e=\dfrac{\sqrt{a^2-b^2}}{a}.$

(18.) In the cubical parabola, whose equation is $y=\dfrac{x^3}{3a^2}$; show that the radius of curvature is $-\dfrac{(a^4+x^4)^{\frac{3}{2}}}{2a^4x}.$

(19.) Prove that in the circle, parabola, ellipse, and hyperbola, or in any plane curve whose equation is of the second degree, the radius of curvature varies as the cube of the normal.

(20.) The equation to the rectangular hyperbola is $y^2 - x^2 + a^2 = 0$; show that the radius of curvature is $\dfrac{(2x^2 - a^2)^{\frac{3}{2}}}{a^2}$, and that the equation to its evolute is $\alpha^{\frac{2}{3}} - \beta^{\frac{2}{3}} = (2a)^{\frac{2}{3}}$.

(21.) Determine the radius of curvature to the curve called the tractrix, the equation being $y = \dfrac{dx}{dy}\sqrt{c^2 - y^2}$.

$$R = \frac{c(c^2 - y^2)^{\frac{1}{2}}}{y}.$$

(22.) The polar equation to the lemniscata of Bernouilli is $r^2 = a^2 \cos 2\theta$; show that the radius of curvature is $\dfrac{a^2}{3r}$.

(23.) Prove that the length of the arc of the evolute intercepted between two radii of curvature is equal to the difference between the lengths of those radii.

(24.) Show that in the common parabola, whose equation is $y^2 = 4ax$, the radius of curvature is greatest at the vertex, that the radius of curvature at that point is half the latus rectum, and determine the equation to the evolute.

(25.) If N be the normal and R the radius of curvature to a point in the ellipse; prove that $N^3 a^2 + R b^4 = 0$.

(26.) $r = \dfrac{a}{\sqrt{\theta}}$ being the equation to the lituus; show that the radius of curvature is $\dfrac{r(4a^4 + r^4)^{\frac{3}{2}}}{2a^2(4a^4 - r^4)}$.

(27.) If $r = f(\theta)$, find an expression for the radius of curvature, that is, prove that

$$R = r\frac{dr}{dp} = \frac{\left(r^2 + \dfrac{dr^2}{d\theta^2}\right)^{\frac{3}{2}}}{r^2 + 2\dfrac{dr^2}{d\theta^2} - r\dfrac{d^2 r}{d\theta^2}}.$$

(28.) The equation to the logarithmic or equiangular spiral, referred to p and r, is $p=mr$; show that the radius of curvature is $\dfrac{p}{m^2}$, and that to this spiral the evolute is a similar spiral.

(29.) $\dfrac{y^2}{a^2}+\dfrac{y^2}{l^2}=1$ being the equation to the ellipse; show that the equation to its evolute is $(a\alpha)^{\frac{2}{3}}+(b\beta)^{\frac{2}{3}}=(a^2-l^2)^{\frac{2}{3}}$, and exhibit its form and position with respect to the centre of the ellipse.

(30.) In the hyperbola, the focus being considered as the pole, the length of the perpendicular on the tangent is $\dfrac{br^{\frac{1}{2}}}{(2a+r)^{\frac{1}{2}}}$; show that the chord of curvature through the focus is $\dfrac{2r(2a+r)}{a}$.

(31.) The equation between p and r in the epicycloid is $(c^2-a^2)p^2=c^2(r^2-a^2)$; prove that the radius of curvature is $\dfrac{1}{c}\sqrt{(c^2-a^2)(r^2-a^2)}$.

(32.) The equation to the involute of the circle is $a\theta+a\sec^{-1}\left(\dfrac{r}{a}\right)=(r^2-a^2)^{\frac{1}{2}}$; prove that its radius of curvature is p, and that its evolute is a circle whose centre is the origin, and radius a.

(33.) The equation to the hypocycloid is $x^{\frac{2}{3}}+y^{\frac{2}{3}}=a^{\frac{2}{3}}$; show that the equation to its evolute is

$$(\alpha+\beta)^{\frac{2}{3}}+(\alpha-\beta)^{\frac{2}{3}}=2a^{\frac{2}{3}}.$$

(34.) Referring to example 22, and letting R and R, respectively represent the radii of curvature at the extremities,

o

of the major and minor axes of an ellipse, prove that the length of the evolute is $4\left(\dfrac{a^2}{b}-\dfrac{b^2}{a}\right)$.

(35.) R being the radius of curvature, and s the length of an arc of a plane curve ; show that $R=\pm\dfrac{ds^3}{dx\,d^2y}$.

(36.) Considering the earth to be an oblate spheroid, or ellipsoid, $2a$ its equatorial and $2b$ its polar diameter, m and m, respectively the lengths of an arc of $1°$ of a meridian in two given latitudes λ and $\lambda_,$, and considering these lengths to coincide with the osculating circles through their middle points ; show, by reference to Ex. 14, that the

equatorial diameter : polar diameter

$$:: \left\{m^{\frac{2}{3}}\sin^2\lambda-m_,^{\frac{2}{3}}\sin^2\lambda_,\right\}^{\frac{1}{2}} : \left\{m_,^{\frac{2}{3}}\cos^2\lambda_,-m^{\frac{2}{3}}\cos^2\lambda\right\}^{\frac{1}{2}}.$$

(37.) Show how the result of the last example would be modified if one of the arcs of the meridian were measured at the equator.

(38.) Let AP be a parabola, P any point in the curve, draw the tangent PT, and the normal PG ; through T, the point in which the tangent intersects the axis of abscissæ, draw TQ at right-angles to that axis, produce PG to meet TQ in Q ; prove that the radius of curvature at P is equal to GQ, and show the centre of the osculating circle.

(39.) The equation to a curve being $x-\sec 2y=0$; show that $\dfrac{1}{p}=2x(x^2-1)^{\frac{1}{2}}$, and that the radius of curvature is $\dfrac{(2x^2-1)^2}{4x}$.

(40.) If, in the common parabola, a point, determined by $x=3a$, be taken ; show that the part of the radius of curvature below the axis of x is $12a$.

(41.) If ds represent the small arc between two points (x, y), $(x+dx, y+dy)$, in a curve, and R the radius of curvature, investigate a general expression for that radius, whatever be the independent variable; that is, prove that

$$R = \frac{ds^3}{d^2x\,dy - d^2y\,dx},$$ and thence deduce expressions for R

when. x, y and s be severally taken as the independent variable.

(42.) Show that, if an inextensible thread were applied to the evolute of a curve, and were to be gradually unwound, a fixed point in the thread would describe the involute or original curve.

(43.) Prove that the tangent to the evolute is the normal to the involute.

(44.) Prove that, when the radius of curvature is either a maximum or a minimum, the contact is of the third order.

CHAPTER XVI.

ENVELOPES TO LINES AND SURFACES.

Considering the evolute to a curve to be generated by the ultimate intersections of consecutive normals, the evolute is their envelope.

If $f(x, y, \alpha) = 0$ be the equation to a system of known curves, intersecting each other in points determined by x and y remaining constant whilst the variable parameter α undergoes an infinitely small variation so as to become $d\alpha$, the problem of finding the equation to the envelope resolves itself into that of finding an equation involving x, y and constant quantities only, α being eliminated between the equations $f(x, y, \alpha)=0$, and $f(x, y, \alpha+d\alpha)=0$.

If there are several equations of condition involving the parameter, it is expedient to have recourse to the method of indeterminate multipliers, as in example 2.

This method of finding envelopes may be applied ·to the determining of the equation to the evolute of a curve.

Ex. (1.) A series of equal ellipses are so placed that their axes are in the same straight lines, the ellipticities alone being variable ; find the equation to the curve which will touch all the ellipses.

Let the constant rectangle $ab=m^2$,

$$\frac{x^2}{a^2}+\frac{y^2}{b^2}=1, \qquad \text{the equation to the ellipse.}$$

Here, a and b being variable, we must consider x, y and m constant, and differentiate with respect to a and b.

$$-\frac{x^2 \cdot 2a}{a^4}-\frac{y^2 \cdot 2b}{b^4}\cdot\frac{db}{da}=0, \quad \frac{y^2}{b^3}\cdot\frac{db}{da}=-\frac{x^2}{a^3}, \quad \therefore \frac{db}{da}=-\frac{b^3 x^2}{a^3 y^2}.$$

$$a\frac{db}{da}+b=0, \qquad \therefore \frac{db}{da}=-\frac{b}{a}.$$

Hence $\frac{b^3 x^2}{a^3 y^2}=\frac{b}{a}$, $b^2 x^2=a^2 y^2$, $\frac{x^2}{a^2}=\frac{y^2}{b^2}$.

$$\left.\begin{array}{c}\therefore \dfrac{x^2}{a^2}+\dfrac{y^2}{b^2}=\dfrac{2y^2}{b^2}=1\\[2mm]=\dfrac{2x^2}{a^2}=1\end{array}\right\} \quad \therefore \frac{4x^2 y^2}{a^2 b^2}=1,$$

$\therefore 2xy=ab=m^2$, the equation to a rectangular hyperbola referred to its asymptotes.

(2.) A straight line, whose length is l, slides down between two rectangular axes x and y ; find the equation to the envelope of the line in all its positions, that is to the curve to which the line is always a tangent.

Let a and b be the variable intercepts of the line on the axes, then

$$\frac{x}{a}+\frac{y}{b}=1, \qquad \text{the equation to the line,}$$

$$a^2+b^2=l^2. \qquad\qquad \text{Euc. b. i. p. 47.}$$

Now, a and b being variable, we must differentiate considering x, y and l constant.

$$-\frac{x}{a^2}\cdot\frac{da}{db}-\frac{y}{b^2}=0, \qquad \frac{x}{a^2}da+\frac{y}{b^2}db=0, \quad . \; . \;\; (1)$$

$$2a\frac{da}{db}+2b=0, \qquad ada+bdb=0. \quad . \;\; . \;\; . \;\; (2)$$

Multiply (2) by the indeterminate multiplier λ.

$$\lambda ada+\lambda bdb=0. \qquad\qquad \text{Add equation (1).}$$

$$\left(\frac{x}{a^2}+\lambda a\right)da+\left(\frac{y}{b^2}+\lambda b\right)db=0.$$

Assume $\quad \frac{x}{a^2}+\lambda a=0, \qquad$ and $\quad \frac{y}{b^2}+\lambda b=0, \qquad$ then

$$\left.\begin{array}{l}\dfrac{x}{a}+\lambda a^2=0 \\[2mm] \dfrac{y}{b}+\lambda b^2=0\end{array}\right\} \quad \therefore \frac{x}{a}+\frac{y}{b}+\lambda(a^2+b^2)=0, \qquad \text{or}$$

$$1=-\lambda l^2, \qquad \therefore \lambda=-\frac{1}{l^2}, \qquad \therefore \frac{x}{a^2}=\frac{a}{l^2}, \qquad \frac{y}{b^2}=\frac{b}{l^2},$$

$$\therefore a=l^{\frac{2}{3}}x^{\frac{1}{3}}, \qquad b=l^{\frac{2}{3}}y^{\frac{1}{3}}, \qquad a^2+b^2=l^{\frac{4}{3}}(x^{\frac{2}{3}}+y^{\frac{2}{3}})=l^2.$$

Hence $x^{\frac{2}{3}}+y^{\frac{2}{3}}=l^{\frac{2}{3}}$, the equation to the locus of the ultimate intersections of the line.

(3.) To determine the curve whose tangent cuts off from the axes a constant area.

First, if the axes be rectangular, let a and b be the variable parts cut off, and $m^2=$ the constant area.

$$\frac{x}{a}+\frac{y}{b}=1, \qquad \text{the equation to the line,} \quad . \quad . \quad (1)$$

$$\frac{ab}{2}=m^2, \qquad \text{the area.} \quad . \quad . \quad . \quad (2)$$

Now, differentiating with respect to the variables a and b, considering x, y and m constant, we have from (1)

$$\frac{x}{a^2}+\frac{y}{b^2}\cdot\frac{db}{da}=0, \qquad \therefore \frac{db}{da}=-\frac{b^2x}{a^2y}, \qquad \text{and from (2)}$$

$$b=\frac{2m^2}{a}, \qquad \therefore \frac{db}{da}=-\frac{2m^2}{a^2}.$$

Hence $\dfrac{b^2x}{a^2y}=\dfrac{2m^2}{a^2}, \qquad b^2=\dfrac{2m^2y}{x}, \qquad b=\dfrac{\sqrt{2}\cdot m\sqrt{y}}{\sqrt{x}}.$

$$a=\frac{2m^2}{b}=2m^2\cdot\frac{\sqrt{x}}{\sqrt{2}\cdot m\sqrt{y}}=\frac{\sqrt{2}m\sqrt{x}}{\sqrt{y}}.$$

$$\therefore \frac{x}{a}+\frac{y}{b}=x\cdot\frac{\sqrt{y}}{\sqrt{2}m\sqrt{x}}+y\cdot\frac{\sqrt{x}}{\sqrt{2}m\sqrt{y}}=\frac{\sqrt{xy}}{m\sqrt{2}}+\frac{\sqrt{xy}}{m\sqrt{2}}=1,$$

$$\therefore 2\sqrt{xy}=m\sqrt{2}, \qquad \sqrt{xy}=\frac{m}{\sqrt{2}}, \qquad xy=\frac{m^2}{2},$$

the equation to a rectangular hyperbola, whose asymptotes are the axes of x and y.

Secondly, if the axes be oblique, let them be inclined at any angle α, a and b being the parts cut off, and m^2 the area ; then

$$\frac{ab\sin\alpha}{2}=m^2, \qquad b=\frac{2m^2}{\sin\alpha}\cdot\frac{1}{a}, \qquad \frac{db}{da}=-\frac{2m^2}{\sin\alpha}\cdot\frac{1}{a^2},$$

$$\text{Also} \quad \frac{db}{da}=-\frac{b^2x}{a^2y} \qquad \text{as in the first case.}$$

Hence $\dfrac{b^2 x}{a^2 y}=\dfrac{2m^2}{a^2\sin\alpha}$, $\quad b^2=\dfrac{2m^2 y}{x\sin\alpha}$, $\quad b=\dfrac{\sqrt{2}\,m\,\sqrt{y}}{\sqrt{x}\,\sqrt{\sin\alpha}}$,

$$a=\dfrac{2m^2}{b\sin\alpha}=\dfrac{2m^2\sqrt{x}}{\sqrt{2}m\sqrt{y}\sqrt{\sin\alpha}}=\dfrac{\sqrt{2}m\sqrt{x}}{\sqrt{y}\sqrt{\sin\alpha}}.$$

$$\therefore \dfrac{x}{a}+\dfrac{y}{b}=\dfrac{x\sqrt{y}\sqrt{\sin\alpha}}{\sqrt{2}m\sqrt{x}}+\dfrac{y\sqrt{x}\sqrt{\sin\alpha}}{\sqrt{2}m\sqrt{y}}=\dfrac{2\sqrt{xy\sin\alpha}}{m\sqrt{2}}=1,$$

$\therefore xy=\dfrac{m^2}{2\sin\alpha}$, the equation to a hyperbola whose asymptotes are the oblique axes Ax, Ay.

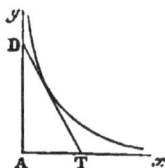

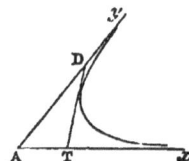

(4.) Determine the equation to the curve which touches all the curves included under the equation

$$y=x\tan\theta-\dfrac{x^2}{4h\cos^2\theta}, \qquad \text{the variable being } \theta.$$

Differentiating with respect to θ, considering x, y and h constant,

$$0=x\cdot\dfrac{1}{\cos^2\theta}-\dfrac{8hx^2\cos\theta\sin\theta}{16h^2\cos^4\theta}, \qquad 1=\dfrac{x}{2h}\cdot\dfrac{\sin\theta}{\cos\theta},$$

$$\tan\theta=\dfrac{2h}{x}, \qquad 1+\tan^2\theta=1+\dfrac{4h^2}{x^2}=\dfrac{x^2+4h^2}{x^2}=\sec^2\theta,$$

$$\therefore \cos^2\theta=\dfrac{x^2}{x^2+4h^2}, \qquad 4h\cos^2\theta=\dfrac{4hx^2}{x^2+4h^2},$$

$$\dfrac{x^2}{4h\cos^2\theta}=x^2\cdot\dfrac{x^2+4h^2}{4hx^2}=\dfrac{x^2}{4h}+h.$$

Hence $y=2h-\dfrac{x^2}{4h}-h=h-\dfrac{x^2}{4h}$, the equation required.

If, in this problem, we consider h to vary as well as θ, and if some constant area $m^2 = h^2 \sin^3\theta \cos\theta$; then we have

$$y = x\tan\theta - \frac{x^2}{4h\cos^2\theta}, \;\cdots (1) \quad \text{and} \quad h = \frac{m}{\sin^{\frac{3}{2}}\theta\cos^{\frac{1}{2}}\theta}, \;\cdots (2)$$

$$\therefore y = x\tan\theta - \frac{x^2}{\dfrac{4m\cos^2\theta}{\sin^{\frac{3}{2}}\theta\cos^{\frac{1}{2}}\theta}} = x\tan\theta - \frac{x^2\sin^{\frac{3}{2}}\theta}{4m\cos^{\frac{3}{2}}\theta},$$

$$\therefore y = x\tan\theta - \frac{x^2}{4m}\tan^{\frac{3}{2}}\theta.$$

Differentiating with respect to θ, considering x, y and m constant,

$$0 = x\sec^2\theta - \frac{x^2}{4m}\cdot\frac{3}{2}\tan^{\frac{1}{2}}\theta\cdot\sec^2\theta, \qquad \frac{3x}{8m}\tan^{\frac{1}{2}}\theta = 1,$$

$$\tan^{\frac{1}{2}}\theta = \frac{8m}{3x}, \qquad \tan\theta = \frac{64m^2}{9x^2}, \qquad \tan^{\frac{3}{2}}\theta = \frac{512m^3}{27x^3}.$$

Whence by substitution in (1) we have

$$y = \frac{64m^2}{9x} - \frac{x^2}{4m}\cdot\frac{512m^3}{27x^2} = \frac{64m^2}{9x} - \frac{128m^2}{27x}$$

$$= \frac{192m^2 - 128m^2}{27x} = \frac{64m^2}{27x}. \qquad \therefore xy = \left(\frac{4}{3}\right)^3 m^2.$$

(5.) Two diameters of a circle intersect at right-angles ; find the locus of the intersections of the chords joining the extremities of the diameters, while the diameters perform a complete revolution.

Let AB, Ab be two semi-diameters at right-angles, $2a$ the diameter of the circle, A the origin of co-ordinates, $r = AP$ the line joining the origin and point of intersection of the chords. Then

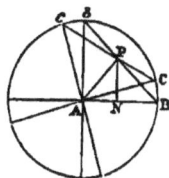

$$\frac{AP}{AB} = \frac{r}{a} = \sin BAP = \sin 45° = \frac{1}{\sqrt{2}}, \qquad \therefore r = \frac{a}{\sqrt{2}}.$$

Now, this problem is the same as that of determining the curve to which the chord at its middle point shall be constantly a tangent; and $y = mx + r\sqrt{m^2+1}$ is the equation to a straight line, r being the perpendicular upon it from the origin.

Differentiating this equation with respect to m, considering x, y, and r constant,

$$0 = x + r\frac{m}{\sqrt{m^2+1}}, \qquad \frac{\sqrt{m^2+1}}{m} = -\frac{r}{x}, \qquad \frac{m^2+1}{m^2} = \frac{r^2}{x^2},$$

$$1 + \frac{1}{m^2} = \frac{r^2}{x^2}, \qquad \frac{1}{m^2} = \frac{r^2}{x^2} - 1 = \frac{r^2-x^2}{x^2}, \qquad m^2 = \frac{x^2}{r^2-x^2},$$

$$m = \frac{x}{\sqrt{r^2-x^2}}, \qquad m^2+1 = \frac{x^2}{r^2-x^2}+1 = \frac{r^2}{r^2-x^2},$$

$$\text{Hence } y = \frac{x^2}{\sqrt{r^2-x^2}} + \frac{r^2}{\sqrt{r^2-x^2}} = \frac{r^2-x^2}{\sqrt{r^2-x^2}} = \sqrt{r^2-x^2},$$

$$\therefore y^2 = r^2 - x^2,$$

$x^2 + y^2 = r^2 = \dfrac{a^2}{2}$, the equation to a circle, whose radius

is $\dfrac{a}{\sqrt{2}}$, and whose centre coincides with that of the original circle.

(6.) If $(x-a)^2 + (y-b)^2 + z^2 = r^2$, and $a^2 + b^2 = c^2$; determine the equation to the envelope of the system of spheres defined by these two equations.

Differentiating with regard to a and b, considering x, y, z and c constant, we have,

$$(x-a)\frac{da}{db} + y - b = 0, \qquad a\frac{da}{db} + b = 0.$$

Multiplying the last equation by the indeterminate multiplier λ, and adding, we have

$$(x+\lambda a-a)\,da+(y+\lambda b-b)\,db=0,$$

$\therefore \lambda a+x-a=0,\;...(1)\qquad \lambda b+y-b=0,\;...(2)\;$; whence by

eliminating λ we have $\dfrac{x}{a}=\dfrac{y}{b}$,

Again (1) $\lambda a^2+ax-a^2=0,$ (2) $\lambda b^2+by-b^2=0,$

$$\therefore \lambda(a^2+b^2)+ax+by-(a^2+b^2)=0,$$

$$\therefore \lambda=1-\frac{ax+by}{c^2},\;...(3)\qquad \because a^2+b^2=c^2.$$

But $\;ax+by=\dfrac{a^2x}{a}+\dfrac{b^2x}{a}=(a^2+b^2)\dfrac{x}{a},\qquad \therefore \dfrac{ax+by}{c^2}=\dfrac{x}{a}.$

Also $x^2+y^2=\dfrac{a^2x^2}{a^2}+\dfrac{b^2x^2}{a^2}=(a^2+b^2)\dfrac{x^2}{a^2},\qquad \therefore \dfrac{(x^2+y^2)^{\frac{1}{2}}}{c}=\dfrac{x}{a}.$

Hence $\;\lambda=1\mp\dfrac{(x^2+y^2)^{\frac{1}{2}}}{c}.$ Substituting in (1), (2),

$$a\pm\frac{a}{c}(x^2+y^2)^{\frac{1}{2}}=-(x-a),\qquad b\pm\frac{b}{c}(x^2+y^2)^{\frac{1}{2}}=-(y-b),$$

$$a^2\pm\frac{2a^2}{c}(x^2+y^2)^{\frac{1}{2}}+\frac{a^2}{c^2}(x^2+y^2)=(x-a)^2,$$

$$b^2\pm\frac{2b^2}{c}(x^2+y^2)^{\frac{1}{2}}+\frac{b^2}{c^2}(x^2+y^2)=(y-b)^2,$$

$$a^2+b^2\pm\frac{2(a^2+b^2)}{c}(x^2+y^2)^{\frac{1}{2}}+\frac{a^2+b^2}{c^2}(x^2+y^2)$$

$$=(x-a)^2+(y-b)^2,\qquad \text{or}$$

$$c^2\pm2c(x^2+y^2)^{\frac{1}{2}}+(x^2+y^2)=r^2-z^2,$$

$\therefore x^2+y^2+z^2\pm2c(x^2+y^2)^{\frac{1}{2}}=r^2-c^2$ is the equation to the envelope of the system of spheres.

(7.) Two straight lines μ and ν, of variable length, are drawn at right-angles to the axis of x, one of them ν passing through the origin of co-ordinates : now if they vary in such a manner that the rectangle contained by them is a constant quantity equal to l^2; determine the curve to which the straight line passing through their upper extremities is always a tangent.

Let $AD=\nu$, $BC=\mu$, $AB=2a$, $AN=x$, $NP=y$. Then

$$BC \cdot AD = \mu\nu = l^2.$$

$$\frac{PN}{TN} = \frac{BC}{TB} = \frac{AD}{AT}, \qquad \text{or}$$

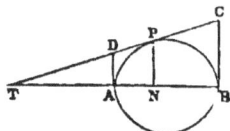

$$\frac{y}{AT+x} = \frac{\mu}{AT+2a} = \frac{\nu}{AT},$$

$$\therefore \; y \cdot AT = \nu \cdot AT + \nu x, \qquad \text{and} \quad y \cdot AT + 2ay = \mu \cdot AT + \mu x,$$

$$(y-\nu)\,AT = \nu x, \qquad\qquad (y-\mu)\,AT = \mu x - 2ay,$$

$$AT = \frac{\nu x}{y-\nu}, \qquad\qquad AT = \frac{\mu x - 2ay}{y-\mu}.$$

Hence $\quad \dfrac{\nu x}{y-\nu} = \dfrac{\mu x - 2ay}{y-\mu}, \qquad\qquad \therefore \; \dfrac{\dfrac{l^2}{\mu} x}{y - \dfrac{l^2}{\mu}} = \dfrac{\mu x - 2ay}{y-\mu},$ or

$$\frac{l^2 x}{\mu y - l^2} = \frac{\mu x - 2ay}{y-\mu}, \qquad \text{or} \quad \mu^2 x = \mu \cdot 2ay + l^2(x-2a),$$

where μ alone is to be considered variable.

Differentiating with respect to μ, we have

$$2\mu x = 2ay, \qquad \therefore \; \mu = \frac{ay}{x}, \qquad \mu^2 = \frac{a^2 y^2}{x^2}.$$

Hence, by substitution, $\quad \dfrac{a^2 y^2}{x} = \dfrac{2a^2 y^2}{x} + l^2(x-2a),$

$$a^2 y^2 = l^2(2ax - x^2), \qquad \text{or}$$

$y^2 = \dfrac{l^2}{a^2}(2ax - x^2)$, the equation to an ellipse, referred to the vertex.

(8.) If a series of parabolas be included under the equation $y^2 = a(x-a)$, a being the variable parameter; show that they will all be touched by the two straight lines determined by the equations $y = +\dfrac{1}{2}x$, $y = -\dfrac{1}{2}x$, and draw these lines.

(9.) Show how the method of determining envelopes may be applied to finding the evolute of a curve, and apply it to determine the evolute of the ellipse, whose equation, referred to the centre, is $\dfrac{x^2}{a^2} + \dfrac{y^2}{b^2} = 1$.

Equation to evolute $(a\alpha)^{\frac{2}{3}} + (b\beta)^{\frac{2}{3}} = (a^2 - b^2)^{\frac{2}{3}}$.

(10.) Prove that the curve which touches all the straight' lines determined by the equation $y = ax + \dfrac{m}{a}$, where a is variable, is the common parabola.

(11.) A system of ellipses, with coincident but variable axes, is subject to the condition that $a^2 + b^2 = m^2$, a and b being the major and minor axes; determine the curve which shall be the envelope of the system.

(12.) If shot be discharged from a cannon with a constant velocity, but at various angles of elevation, they will describe the parabolas included under the equation $y = ax - (1 + a^2)\dfrac{x^2}{4c}$, a being the variable parameter. Show that the curve which will touch all these parabolas is itself a parabola whose equation is $y = c - \dfrac{x^2}{4c}$.

(13.) Considering the envelope to be formed by the intersections of straight lines; show that the problem " to determine the equation to the envelope" is the inverse of the problem " to determine the equation to a tangent to a curve."

(14.) If p_i be a perpendicular of constant length from the origin upon the straight lines defined by $y=ax+p_i(a^2+1)^{\frac{1}{2}}$; show that the envelope of all these lines is a circle whose radius is p_i.

(15.) If a surface be produced by the continued intersection of planes represented by the equation $\frac{x}{a}+\frac{y}{b}+\frac{z}{c}=1$, where $abc=m^3$; a, b, c being variable, and m^3 constant; prove that the equation to the surface is $xyz=\left(\frac{m}{3}\right)^3$.

(16.) A straight line, cutting from two straight lines which meet in any angle, two segments whose sum is a, is a tangent to a curve; prove that that curve is a parabola, and trace it.

(17.) If on one side of a horizontal straight line AR an indefinite number of parabolas of equal area be described from a common point A, with their axes perpendicular to AR, the equation to this system of parabolas is $ay=2a^{\frac{1}{3}}a^{\frac{2}{3}}x-x^2$, where α is variable; prove that the curve which will touch them all is an equilateral hyperbola whose equation is $xy=\frac{2^5}{3^3}a^2$, AR and a perpendicular to it from A being its asymptotic axes.

CHAPTER XVII.

(1.) Prove the ordinary rules for differentiation.

(2.) Explain the difference between explicit and implicit functions.

(3.) Define and illustrate the terms " limit," " differential," " differential coefficient."

(4.) Explain the difference between algebraic and transcendental functions.

(5.) Investigate the differentials of $u=\sin x$, $u=n\tan\theta$, $u=a^x$, $u=\log x$.

(6.) Prove Taylor's Theorem, and from it deduce Stirling's or Maclaurin's Theorem, and the Binomial Theorem of Newton.

(7.) If $y = e^x\sin x$; show, by means of the theorem of Leibnitz, that $\dfrac{d^n y}{dx^n}=2^{\frac{n}{2}} e^x\sin\left(x+n\dfrac{\pi}{4}\right)$.

(8.) In what manner may the value of a fraction be determined when its numerator and denominator vanish simultaneously?

(9.) If $u=f(x)$; show that u is a maximum or minimum when an odd number of differential coefficients becoming$=0$, the differential coefficient of the next succeeding order is negative or positive.

(10.) Deduce the equation to a straight line, $y=mx+b$, and show that the equation to a perpendicular to it is $y=-\dfrac{1}{m}x+b$.

(11.) Show that the equation to a straight line, which intersects the axis of x at a distance a from the origin of co-ordinates, and the axis of y at a distance b from that origin, is $\frac{y}{b} + \frac{x}{a} = 1$.

(12.) Show that the equation to a tangent to a curve, referred to rectangular co-ordinates, is $(y, -y) = \frac{dy}{dx} (x, -x)$.

(13.) If AT and AD be the intercepts of the tangent on the axes of x and y respectively; prove that $AT = y \frac{dx}{dy} - x$, and $AD = y - x \frac{dy}{dx}$, and determine the equation to the normal.

(14.) Determine the differential expression for the subtangent, subnormal, tangent, normal, perpendicular on tangent, and the tangent of the angle which the tangent makes with a line from the origin.

(15.) If $u = f(x, y)$; prove that $du = \left(\frac{du}{dx}\right) dx + \left(\frac{du}{dy}\right) dy$, and that $\frac{d^2u}{dy\,dx} = \frac{d^2u}{dx\,dy}$.

(16.) If $u = f(y, z)$, where y, z, and consequently u, are functions of x; show that $du = \left(\frac{du}{dy}\right) dy + \left(\frac{du}{dz}\right) dz$.

(17.) Determine the conditions upon which a function of two independent variables is a maximum or minimum.

(18.) Determine the differential expression for the area of a plane curve, and if s be the length, and $\frac{dy}{dx} = p$; prove that $\frac{ds}{dx} = (1 + p^2)^{\frac{1}{2}}$.

(19.) If S be the surface and V the volume of a solid generated by the revolution of a curve round its axis ; show that $\dfrac{dV}{dx}=\pi y^2$,   and $\dfrac{dS}{dx}=2\pi y^2(1+p^2)^{\frac{1}{2}}$.

(20.) If r and r, be the radii of the greater and smaller ends of the frustrum of a right cone, and a the slant height; prove that the area of the frustrum is $\pi a\,(r+r,)$.

(21.) If r be the radius vector, p the perpendicular on the tangent, and θ the angle swept out by the revolution of r round the pole S; show that $\dfrac{1}{p^2}=u^2+\left(\dfrac{du}{d\theta}\right)^2$, where $u=\dfrac{1}{r}$; and that $\dfrac{d\theta}{dr}=\dfrac{p}{r\,(r^2-p^2)^{\frac{1}{2}}}$.

(22.) If in polar curves p be the length of the perpendicular upon the tangent ; find the value of p in the circle, parabola, ellipse, and hyperbola.

(23.) Define the rectilinear asymptote and the asymptotic circle.

(24.) Define conjugate points, double points, cusps, and points of contrary flexure, and show that a curve is concave or convex to the axis according as y and $\dfrac{d^2y}{dx^2}$ have the same or different signs.

(25.) Prove that, in spirals, the curve is concave or convex towards the pole, according as $\dfrac{dp}{dr}$ is positive or negative.

(26.) If A be the area, and s the length of a plane curve ; prove that $\dfrac{dA}{dx}=y$,   and $\dfrac{dA}{d\theta}=\dfrac{1}{2}r^2$, $\dfrac{ds}{d\theta}=\sqrt{r^2+\left(\dfrac{dr}{d\theta}\right)^2}$ and $\dfrac{ds}{dr}=\dfrac{r}{(r^2-p^2)^{\frac{1}{2}}}$.

(27.) Prove that, in spirals, the subtangent $= r^2 \dfrac{d\theta}{dr}$

$= \dfrac{pr}{(r^2 - p^2)^{\frac{1}{2}}}$, and show how to draw the asymptote to a spiral.

(28.) Explain what is meant by the osculating circle; and show that the evolutes of all algebraic curves are rectificable.

(29.) Explain the theory of the different orders of contact of plane curves; point out the exceptions to the rule that every curve is cut by its circle of curvature, and show how these exceptions apply to the ellipse.

(30.) Explain the difference between Taylor's and Maclaurin's Theorems, and point out the circumstances under which the former sometimes fails.

(31.) Investigate Lagrange's[*] Theorem, and apply it to determine a general law for the inversion of series by means of the equation $x = ay + by^2 + cy^3 + dy^4 + \&\text{c}.$

(32.) Apply Lagrange's Theorem to the determination of the four first terms of the development of y^m, when $y = a + xy^n$; and find the general term in the expansion of x^m in a series of powers of $\cos\theta$, when $x + \dfrac{1}{x} = 2\cos\theta.$

(33.) If $u = \dfrac{d^2y}{dx^2} - \left(x\dfrac{dy}{dx} - y \right) \cdot \dfrac{1}{1 - x^2}$, x being the inde-

---

* If $y = z + x\phi(y)$, and if $u = f(y)$, f and ϕ being any functions whatever, then

$$u = f(z) + \left[\phi(z) f'(z) \right] \dfrac{x}{1} + \dfrac{d}{dz} \left[\{\phi(z)\}^2 f'(z) \right] \dfrac{x^2}{1 \cdot 2}$$

$$+ \dfrac{d^2}{dz^2} \left[\{\phi(z)\}^3 f'(z) \right] \dfrac{x^3}{1 \cdot 2 \cdot 3} + \dfrac{d^3}{dz^3} \left[\{\phi(z)\}^4 f'(z) \right] \dfrac{x^4}{1 \cdot 2 \cdot 3 \cdot 4} + \&\text{c}.$$

pendent variable ; show that, when x becomes $\cos\theta$, and θ is made the independent variable, $u=\left(\dfrac{d^2y}{d\theta^2}+y\right)\operatorname{cosec}^2\theta$.

(34.) Explain exactly the mode in which the following curves are generated, construct them, and thence derive their equations : namely, the circle, parabola, ellipse, hyperbola, cissoid of Diocles, conchoid of Nicomedes (superior and inferior), cycloid, epicycloid, lemniscata of Bernouilli, quadratrix of Dinostratus, involute of the circle, catenary, tractory, elastic curve, witch of Agnesi, curve of sines, cardioid, trisectrix, logarithmic or equiangular spiral, spiral of Archimedes, hyperbolic or reciprocal spiral, lituus, parabolic spiral.

(35.) Show what kind of curves are included under the equations $y^2=mx+nx^2$, $r=a\sin n\theta$, $r=a\cos\theta+b$, $r=a\theta^n$, $r=a\sin n\theta+b\sin n,\theta+c\sin n,,\theta+$ &c. respectively.

PRINTED BY COX (BROTHERS) AND WYMAN, GREAT QUEEN STREET.

7, *Stationers' Hall Court,*
March, 1872.

1862
LONDINI
HONORIS
CAUSA

NEW LIST

OF

WEALE'S

RUDIMENTARY, SCIENTIFIC, EDUCATIONAL, AND CLASSICAL SERIES,

OF WORKS SUITABLE FOR

Engineers, Architects, Builders, Artisans, and Students generally, as well as to those interested in Workmen's Libraries, Free Libraries, Literary and Scientific Institutions, Colleges, Schools, Science Classes, &c., &c.

** THE ENTIRE SERIES IS FREELY ILLUSTRATED WHERE REQUISITE.

(*The Volumes contained in this List are bound in limp cloth, except where otherwise stated.*)

AGRICULTURE.

66. CLAY LANDS AND LOAMY SOILS, by J. Donaldson. 1s.

140. SOILS, MANURES, AND CROPS, by R. Scott Burn. 2s.

141. FARMING, AND FARMING ECONOMY, Historical and Practical, by R. Scott Burn. 3s.

142. CATTLE, SHEEP, AND HORSES, by R. Scott Burn. 2s. 6d.

145. MANAGEMENT OF THE DAIRY—PIGS—POULTRY, by R. Scott Burn. With Notes on the Diseases of Stock. 2s.

146. UTILISATION OF TOWN SEWAGE—IRRIGATION—RECLAMATION OF WASTE LAND, by R. Scott Burn. 2s. 6d.

Nos. 140, 141, 142, 145, *and* 146 *bound in 2 vols., cloth boards,* 14s.

CULTURE OF FRUIT TREES, by De Breuil. 191 Woodcuts. [*Just ready.*

LOCKWOOD & CO., 7, STATIONERS' HALL COURT.

ARCHITECTURE AND BUILDING.

16. ARCHITECTURE, Orders of, by W. H. Leeds. 1s. 6d. } In 1

17. ——————— Styles of, by T. Talbot Bury. 1s. 6d. } vol., 2s. 6d.

18. ——————— Principles of Design, by E. L. Garbett. 2s.
    Nos. 16, 17, *and* 18 *in* 1 *vol. cloth boards,* 5s. 6d.

22. BUILDING, the Art of, by E. Dobson. 1s. 6d.

23. BRICK AND TILE MAKING, by E. Dobson. 3s.

25. MASONRY AND STONE-CUTTING, by E. Dobson. New Edition, with Appendix on the Preservation of Stone. 2s. 6d.

30. DRAINAGE AND SEWAGE OF TOWNS AND BUILD-INGS, by G. D. Dempsey. 2s.
    With No. 29 (*See page* 4), *Drainage of Districts and Lands,* 3s.

35. BLASTING AND QUARRYING OF STONE, &c., by Field-Marshal Sir J. F. Burgoyne. 1s. 6d.

36. DICTIONARY OF TECHNICAL TERMS used by Architects, Builders, Engineers, Surveyors, &c. New Edition, revised and enlarged by Robert Hunt, F.G.S. [*In preparation.*

42. COTTAGE BUILDING, by C. B. Allen. 1s.

44. FOUNDATIONS & CONCRETE WORKS, by Dobson. 1s. 6d.

45. LIMES, CEMENTS, MORTARS, &c., by Burnell. 1s. 6d.

57. WARMING AND VENTILATION, by C. Tomlinson, F.R.S. 3s

83**. DOOR LOCKS AND IRON SAFES, by Tomlinson. 2s. 6d.

111. ARCHES, PIERS, AND BUTTRESSES, by W. Bland. 1s. 6d.

116. ACOUSTICS OF PUBLIC BUILDINGS, by T.R. Smith. 1s. 6d.

123. CARPENTRY AND JOINERY, founded on Robison and Tredgold. 1s. 6d.

123*. ILLUSTRATIVE PLATES to the preceding. 4to. 4s. 6d.

124. ROOFS FOR PUBLIC AND PRIVATE BUILDINGS, founded on Robison, Price, and Tredgold. 1s. 6d.

124*. PLATES OF RECENT IRON ROOFS. 4to. [*Reprinting.*

127. ARCHITECTURAL MODELLING IN PAPER, Practical Instructions, by T. A. Richardson, Architect. 1s. 6d.

128. VITRUVIUS'S ARCHITECTURE, by J. Gwilt, Plates. 5s.

130. GRECIAN ARCHITECTURE, Principles of Beauty in, by the Earl of Aberdeen. 1s.
    Nos. 128 *and* 130 *in* 1 *vol. cloth boards,* 7s.

132. ERECTION OF DWELLING-HOUSES, with Specifications, Quantities of Materials, &c., by S. H. Brooks, 27 Plates. 2s. 6d.

156. QUANTITIES AND MEASUREMENTS, by Beaton. 1s. 6d.

175. BUILDERS' AND CONTRACTORS' PRICE-BOOK, by G. R. Burnell. 3s. 6d.   [*Now ready.*

PUBLISHED BY LOCKWOOD & CO.,

ARITHMETIC AND MATHEMATICS.

32. MATHEMATICAL INSTRUMENTS, THEIR CONSTRUC-
TION, USE, &c., by J. F. Heather. Original Edition in
1 vol. 1s. 6d.
₊ In ordering the above, be careful to say "Original Edition," to
distinguish it from the Enlarged Edition in 3 vols., advertised
on page 4 as now ready.
60. LAND AND ENGINEERING SURVEYING, by T. Baker. 2s.
61*. READY RECKONER for the Admeasurement and Valuation
of Land, by A. Arman. 1s. 6d.
76. GEOMETRY, DESCRIPTIVE, with a Theory of Shadows and
Perspective, and a Description of the Principles and Practice
of Isometrical Projection, by J. F. Heather. 2s.
83. COMMERCIAL BOOK-KEEPING, by James Haddon. 1s.
84. ARITHMETIC, with numerous Examples, by J. R. Young. 1s. 6d.
84*. KEY TO THE ABOVE, by J. R. Young. 1s. 6d.
85. EQUATIONAL ARITHMETIC: including Tables for the
Calculation of Simple Interest, with Logarithms for Compound
Interest, and Annuities, by W. Hipsley. 1s.
85*. SUPPLEMENT TO THE ABOVE, 1s.
85 and 85* in 1 vol., 2s.
86. ALGEBRA, by J. Haddon. 2s.
86*. KEY AND COMPANION to the above, by J. R. Young. 1s. 6d.
88. THE ELEMENTS OF EUCLID, with Additional Propositions,
and Essay on Logic, by H. Law. 2s.
90. ANALYTICAL GEOMETRY AND CONIC SECTIONS, by
J. Hann. Entirely New Edition, improved and re-written
by J. R. Young. 2s. [Now ready.
91. PLANE TRIGONOMETRY, by J. Hann. 1s.
92. SPHERICAL TRIGONOMETRY, by J. Hann. 1s.
Nos. 91 and 92 in 1 vol., 2s.
93. MENSURATION, by T. Baker. 1s. 6d.
94. MATHEMATICAL TABLES, LOGARITHMS, with Tables of
Natural Sines, Cosines, and Tangents, by H. Law, C.E. 2s. 6d.
101. DIFFERENTIAL CALCULUS, by W. S. B. Woolhouse. 1s.
101*. WEIGHTS, MEASURES, AND MONEYS OF ALL
NATIONS; with the Principles which determine the Rate of
Exchange, by W. S. B. Woolhouse. 1s. 6d.
102. INTEGRAL CALCULUS, RUDIMENTS, by H. Cox, B.A. 1s.
103. INTEGRAL CALCULUS, Examples on, by J. Hann. 1s.
104. DIFFERENTIAL CALCULUS, Examples, by J. Haddon. 1s.
105. ALGEBRA, GEOMETRY, and TRIGONOMETRY, in Easy
Mnemonical Lessons, by the Rev. T. P. Kirkman. 1s. 6d.
117. SUBTERRANEOUS SURVEYING, AND THE MAG-
NETIC VARIATION OF THE NEEDLE, by T. Fenwick,
with Additions by T. Baker. 2s. 6d.

7, STATIONERS' HALL COURT, LUDGATE HILL.

131. READY-RECKONER FOR MILLERS, FARMERS, AND MERCHANTS, showing the Value of any Quantity of Corn, with the Approximate Values of Mill-stones & Mill Work. 1s.

136. RUDIMENTARY ARITHMETIC, by J. Haddon, edited by A. Arman. 1s. 6d.

137. KEY TO THE ABOVE, by A. Arman. 1s. 6d. ·

147. STEPPING STONE TO ARITHMETIC, by A. Arman. 1s.

148. KEY TO THE ABOVE, by A. Arman. 1s.

158. THE SLIDE RULE, AND HOW TO USE IT. With Slide Rule in a pocket of cover. 3s.

168. DRAWING AND MEASURING INSTRUMENTS. Including—Instruments employed in Geometrical and Mechanical Drawing, the Construction, Copying, and Measurement of Maps, Plans, &c., by J. F. HEATHER, M.A. 1s. 6d.
[Now ready.

169. OPTICAL INSTRUMENTS, more especially Telescopes, Microscopes, and Apparatus for producing copies of Maps and Plans by Photography, by J. F. HEATHER, M.A. 1s. 6d.
[Now ready.

170. SURVEYING AND ASTRONOMICAL INSTRUMENTS. Including—Instruments Used for Determining the Geometrical Features of a portion of Ground, and in Astronomical Observations, by J. F. HEATHER, M.A. 1s. 6d. [Now ready.

⁎ The above three volumes form an enlargement of the Author's original work, " Mathematical Instruments," the Tenth Edition of which (No. 32) is still on sale, price 1s. 6d.

⁎ New Volumes in preparation :—

PRACTICAL PLANE GEOMETRY : Giving the Simplest Modes of Constructing Figures contained in one Plane, by J. F. HEATHER, M.A.

PROJECTION, Orthographic, Topographic, and Perspective : giving the various modes of Delineating Solid Forms by Constructions on a Single Plane Surface, by J. F. HEATHER, M.A.

⁎ The above two volumes, with the Author's work already in the Series, " Descriptive Geometry," will form a complete Elementary Course of Mathematical Drawing.

CIVIL ENGINEERING.

13. CIVIL ENGINEERING, by H. Law and G. R. Burnell. Fifth Edition, with Additions. 5s.

29. DRAINAGE OF DISTRICTS AND LANDS, by G. D. Dempsey. 1s. 6d.
With No. 30 (See page 2), Drainage and Sewage of Towns, 4s.

31. WELL-SINKING, BORING, AND PUMP WORK, by J. G.
Swindell, revised by G. R. Burnell. 1s.
43. TUBULAR AND IRON GIRDER BRIDGES, including the
Britannia and Conway Bridges, by G. D. Dempsey. 1s. 6d.
46. ROAD-MAKING AND MAINTENANCE OF MACADA-
MISED ROADS, by Field-Marshal Sir J. F. Burgoyne. 1s. 6d
47. LIGHTHOUSES, their Construction and Illumination, by Alan
Stevenson. 3s.
62. RAILWAY CONSTRUCTION, by Sir M. Stephenson. With
Additions by E. Nugent, C.E. 2s. 6d.
62*. RAILWAY CAPITAL AND DIVIDENDS, with Statistics of
Working, by E. D. Chattaway. 1s.
No. 62 and 62* in 1 vol., 3s. 6d.
80*. EMBANKING LANDS FROM THE SEA, by J. Wiggins. 2s.
82**. GAS WORKS, and the PRACTICE of MANUFACTURING
and DISTRIBUTING COAL GAS, by S. Hughes. 3s.
82***. WATER-WORKS FOR THE SUPPLY OF CITIES AND
TOWNS, by S. Hughes, C.E. 3s.
118. CIVIL ENGINEERING OF NORTH AMERICA, by D.
Stevenson. 3s.
120. HYDRAULIC ENGINEERING, by G. R. Burnell. 3s.
121. RIVERS AND TORRENTS, with the Method of Regulating
their Course and Channels, Navigable Canals, &c., from the
Italian of Paul Frisi. 2s. 6d.

EMIGRATION.

154. GENERAL HINTS TO EMIGRANTS. 2s.
157. EMIGRANT'S GUIDE TO NATAL, by R. J. Mann, M.D. 2s.
159. EMIGRANT'S GUIDE TO NEW SOUTH WALES,
WESTERN AUSTRALIA, SOUTH AUSTRALIA, VIC-
TORIA, AND QUEENSLAND, by James Baird, B.A. 2s. 6d.
160. EMIGRANT'S GUIDE TO TASMANIA AND NEW ZEA-
LAND, by James Baird, B.A. 2s. [*Ready.*

FINE ARTS.

20. PERSPECTIVE, by George Pyne. 2s.
27. PAINTING; or, A GRAMMAR OF COLOURING, by G.
Field. 2s.
40. GLASS STAINING, by Dr. M. A. Gessert, with an Appendix
on the Art of Enamel Painting, &c. 1s.
41. PAINTING ON GLASS, from the German of Fromberg. 1s.
69. MUSIC, Treatise on, by C. C. Spencer. 2s.
71. THE ART OF PLAYING THE PIANOFORTE, by C. C.
Spencer. 1s.

LEGAL TREATISES.

50. LAW OF CONTRACTS FOR WORKS AND SERVICES, by David Gibbons. 1s. 6d.

107. THE COUNTY COURT GUIDE, by a Barrister. 1s. 6d.

108. METROPOLIS LOCAL MANAGEMENT ACTS. 1s. 6d.

108*. METROPOLIS LOCAL MANAGEMENT AMENDMENT ACT, 1862; with Notes and Index. 1s.
Nos. 108 *and* 108* *in* 1 *vol.*, 2s. 6d.

109. NUISANCES REMOVAL AND DISEASES PREVENTION AMENDMENT ACT. 1s.

110. RECENT LEGISLATIVE ACTS applying to Contractors, Merchants, and Tradesmen. 1s.

151. THE LAW OF FRIENDLY, PROVIDENT, BUILDING, AND LOAN SOCIETIES, by N. White. 1s.

163. THE LAW OF PATENTS FOR INVENTIONS, by F. W. Campin, Barrister. 2s.

MECHANICS & MECHANICAL ENGINEERING.

6. MECHANICS, by Charles Tomlinson. 1s. 6d.

12. PNEUMATICS, by Charles Tomlinson. New Edition. 1s. 6d.

33. CRANES AND MACHINERY FOR RAISING HEAVY BODIES, the Art of Constructing, by J. Glynn. 1s.

34. STEAM ENGINE, by Dr. Lardner. 1s.

59. STEAM BOILERS, their Construction and Management, by R. Armstrong.. With Additions by R. Mallet. 1s. 6d.

63. AGRICULTURAL ENGINEERING, BUILDINGS, MOTIVE POWERS, FIELD MACHINES, MACHINERY AND IMPLEMENTS, by G. H. Andrews, C.E. 3s.

67. CLOCKS, WATCHES, AND BELLS, by E. B. Denison. New Edition, with Appendix. 3s. 6d.
Appendix (to the 4th and 5th Editions) separately, 1s.

77*. ECONOMY OF FUEL, by T. S. Prideaux. 1s. 6d.

78. STEAM AND LOCOMOTION, by Sewell. [*Reprinting.*

78*. THE LOCOMOTIVE ENGINE, by G. D. Dempsey. 1s. 6d.

79*. ILLUSTRATIONS TO ABOVE. 4to. 4s. 6d. [*Reprinting.*

80. MARINE ENGINES, AND STEAM VESSELS, AND THE SCREW, by Robert Murray, C.E., Engineer Surveyor to the Board of Trade. With a Glossary of Technical Terms, and their equivalents in French, German, and Spanish. 3s.

82. WATER POWER, as applied to Mills, &c., by J. Glynn. 2s.

97. STATICS AND DYNAMICS, by T. Baker. New Edition. 1s. 6d.

98. MECHANISM AND MACHINE TOOLS, by T. Baker; and TOOLS AND MACHINERY, by J. Nasmyth. 2s. 6d.

113*. MEMOIR ON SWORDS, by Marey, translated by Maxwell. 1s.

PUBLISHED BY LOCKWOOD & CO.,

114. MACHINERY, Construction and Working, by C.D. Abel. 1s.6d.
115. PLATES TO THE ABOVE. 4to. 7s. 6d.
125. COMBUSTION OF COAL, AND THE PREVENTION OF SMOKE, by C. Wye Williams, M.I.C.E. 3s.
139. STEAM ENGINE, Mathematical Theory of, by T. Baker. 1s.
162. THE BRASSFOUNDER'S MANUAL, by W. Graham. 2s. 6d.
164. MODERN WORKSHOP PRACTICE. By J.G. Winton. 3s.
165. IRON AND HEAT, Exhibiting the Principles concerned in the Construction of Iron Beams, Pillars, and Bridge Girders, and the Action of Heat in the Smelting Furnace, by James Armour, C.E. Woodcuts. 2s. 6d. [Now ready.
166. POWER IN MOTION: Horse Power, Motion, Toothed Wheel Gearing, Long and Short Driving Bands, Angular Forces, &c., by James Armour, C.E. With 73 Diagrams. 2s.6d. [Now ready.
167. A TREATISE ON THE CONSTRUCTION OF IRON BRIDGES, GIRDERS, ROOFS, AND OTHER STRUCTURES, by F. Campin. Numerous Woodcuts. 2s. [Ready.
171. THE WORKMAN'S MANUAL OF ENGINEERING DRAWING, by John Maxton, Instructor in Engineering Drawing, Royal School of Naval Architecture & Marine Engineering, South Kensington. Plates & Diagrams. 3s. 6d. [Ready.
172. MINING TOOLS. For the Use of Mine Managers, Agents, Mining Students, &c., by William Morgans, Lecturer on Mining, Bristol School of Mines. 12mo. 2s.6d. [Just ready.
172*. ATLAS OF PLATES to the above, containing 200 Illustrations. 4to. 4s. 6d. [Just ready.
176. TREATISE ON THE METALLURGY OF IRON; containing Outlines of the History of Iron Manufacture, Methods of Assay, and Analysis of Iron Ores, Processes of Manufacture of Iron and Steel, &c., by H. Bauerman, F.G.S., A.R.S.M. Second Edition, revised and enlarged. Woodcuts. 4s.6d. [Ready.
COAL AND COAL MINING, by W.W. Smyth. [In preparation.

NAVIGATION AND SHIP-BUILDING.

51. NAVAL ARCHITECTURE, by J. Peake. 3s.
53*. SHIPS FOR OCEAN AND RIVER SERVICE, Construction of, by Captain H. A. Sommerfeldt. 1s.
53**. ATLAS OF 15 PLATES TO THE ABOVE, Drawn for Practice. 4to. 7s. 6d. [Reprinting.
54. MASTING, MAST-MAKING, and RIGGING OF SHIPS, by R. Kipping. 1s. 6d.
54*. IRON SHIP-BUILDING, by J. Grantham. Fifth Edition, with Supplement. 4s.
54**. ATLAS OF 40 PLATES to illustrate the preceding. 4to. 38s.

55. NAVIGATION ; the Sailor's Sea Book: How to Keep the Log and Work it off, Law of Storms, &c., by J. Greenwood. 2s.
83 bis. SHIPS AND BOATS, Form of, by W. Bland. 1s. 6d.
99. NAUTICAL ASTRONOMY AND NAVIGATION, by J. R. Young. 2s.
100*. NAVIGATION TABLES, for Use with the above. 1s. 6d.
106. SHIPS' ANCHORS for all SERVICES, by G. Cotsell. 1s. 6d.
149. SAILS AND SAIL-MAKING, by R. Kipping, N.A. 2s. 6d.
155. ENGINEER'S GUIDE TO THE ROYAL AND MER-CANTILE NAVIES, by a Practical Engineer. Revised by D. F. McCarthy. 3s.

PHYSICAL AND CHEMICAL SCIENCE.

1. CHEMISTRY, by Prof. Fownes. With Appendix on Agricultural Chemistry. New Edition, with Index. 1s.
2. NATURAL PHILOSOPHY, by Charles Tomlinson. 1s.
3. GEOLOGY, by Major-Gen. Portlock. New Edition. 1s. 6d.
4. MINERALOGY, by A. Ramsay, Jun. 3s.
7. ELECTRICITY, by Sir W. S. Harris. 1s. 6d.
7*. GALVANISM, ANIMAL AND VOLTAIC ELECTRICITY, by Sir W. S. Harris. 1s. 6d.
8. MAGNETISM, by Sir W. S. Harris. New Edition, revised and enlarged by H. M. Noad, Ph.D., F.R.S. With 165 woodcuts. 3s. 6d. [This day.
11. HISTORY AND PROGRESS OF THE ELECTRIC TELE-GRAPH, by Robert Sabine, C.E., F.S.A. 3s.
72. RECENT AND FOSSIL SHELLS (A Manual of the Mollusca), by S. P. Woodward. With Appendix by Ralph Tate, F.G.S. 6s. 6d.; in cloth boards, 7s. 6d. Appendix separately, 1s,
79**. PHOTOGRAPHY, the Stereoscope, &c., from the French of D. Van Monckhoven, by W. H. Thornthwaite. 1s. 6d.
96. ASTRONOMY, by the Rev. R. Main. New and Enlarged Edition, with an Appendix on "Spectrum Analysis." 1s. 6d.
133. METALLURGY OF COPPER, by Dr. R. H. Lamborn. 2s.
134. METALLURGY OF SILVER AND LEAD, by Lamborn. 2s.
135. ELECTRO-METALLURGY, by A. Watt. 2s.
138. HANDBOOK OF THE TELEGRAPH, by R. Bond. New and enlarged Edition. 1s. 6d.
143. EXPERIMENTAL ESSAYS—On the Motion of Camphor and Modern Theory of Dew, by C. Tomlinson. 1s.
161. QUESTIONS ON MAGNETISM, ELECTRICITY, AND PRACTICAL TELEGRAPHY, by W. McGregor. 1s. 6d.
173. PHYSICAL GEOLOGY (partly based on Portlock's "Rudiments of Geology "), by Ralph Tate, A.L.S., &c. 2s. [Now ready.
174. HISTORICAL GEOLOGY (partly based on Portlock's "Rudiments of Geology "), by Ralph Tate, A.L.S., &c. 2s. 6d. [Now ready.

PUBLISHED BY LOCKWOOD & CO.,

MISCELLANEOUS TREATISES.

12. DOMESTIC MEDICINE, by Dr. Ralph Gooding. 2s.
112*. THE MANAGEMENT OF HEALTH, by James Baird. 1s.
113. USE OF FIELD ARTILLERY ON SERVICE, by Taubert, translated by Lieut.-Col. H. H. Maxwell. 1s. 6d.
150. LOGIC, PURE AND APPLIED, by S. H. Emmens. 1s. 6d.
152. PRACTICAL HINTS FOR INVESTING MONEY: with an Explanation of the Mode of Transacting Business on the Stock Exchange, by Francis Playford, Sworn Broker. 1s.
153. LOCKE ON THE CONDUCT OF THE HUMAN UNDERSTANDING, Selections from, by S. H. Emmens. 2s.

NEW SERIES OF EDUCATIONAL WORKS.

1. ENGLAND, History of, by W. D. Hamilton. 5s.; cloth boards, 6s. (Also in 5 parts, price 1s. each.)
5. GREECE, History of, by W. D. Hamilton and E. Levien, M.A. 2s. 6d.; cloth boards, 3s. 6d.
7. ROME, History of, by E. Levien. 2s. 6d.; cloth boards, 3s. 6d.
9. CHRONOLOGY OF HISTORY, ART, LITERATURE, and Progress, from the Creation of the World to the Conclusion of the Franco-German War. The continuation by W. D. Hamilton, F.S.A. 3s. cloth limp; 3s. 6d. cloth boards.
[Now ready.
11. ENGLISH GRAMMAR, by Hyde Clarke, D.C.L. 1s.
11*. HANDBOOK OF COMPARATIVE PHILOLOGY, by Hyde Clarke, D.C.L. 1s.
12. ENGLISH DICTIONARY, containing above 100,000 words, by Hyde Clarke, D.C.L. 3s. 6d.; cloth boards, 4s. 6d.
————————, with Grammar. Cloth bds. 5s. 6d.
14. GREEK GRAMMAR, by H. C. Hamilton. 1s.
15. ———— DICTIONARY, by H. R. Hamilton. Vol. 1. Greek—English. 2s.
17. ———— Vol. 2. English—Greek. 2s.
———— Complete in 1 vol. 4s.; cloth boards, 5s.
———————————, with Grammar. Cloth boards, 6s.
19. LATIN GRAMMAR, by T. Goodwin, M.A. 1s.
20. ———— DICTIONARY, by T. Goodwin, M.A. Vol. 1. Latin—English. 2s.
22. ———— Vol. 2. English—Latin. 1s. 6d.
———— Complete in 1 vol. 3s. 6d.; cloth boards, 4s. 6d.
———————————, with Grammar. Cloth bds. 5s. 6d.
24. FRENCH GRAMMAR, by G. L. Strauss. 1s.

25. FRENCH DICTIONARY, by Elwes. Vol. 1. Fr.—Eng. 1s.
26. ————————— Vol. 2. English—French. 1s. 6d.
    ————— Complete in 1 vol. 2s. 6d.; cloth boards, 3s. 6d.
    —————————————, with Grammar. Cloth bds. 4s. 6d.
27. ITALIAN GRAMMAR, by A. Elwes. 1s.
28. ————— TRIGLOT DICTIONARY, by A. Elwes. Vol. 1.
    Italian—English—French. 2s.
30. ————— Vol. 2. English—French—Italian. 2s.
32. ————— Vol. 3. French—Italian—English. 2s.
    ————— Complete in 1 vol. Cloth boards, 7s. 6d.
    —————————————, with Grammar. Cloth bds. 8s. 6d.
34. SPANISH GRAMMAR, by A. Elwes. 1s.
35. ————— ENGLISH AND ENGLISH—SPANISH DIC-
    TIONARY, by A. Elwes. 4s.; cloth boards, 5s.
    —————————————, with Grammar. Cloth boards, 6s.
39. GERMAN GRAMMAR, by G. L. Strauss. 1s.
40. ————— READER, from best Authors. 1s.
41. ————— TRIGLOT DICTIONARY, by N. E. S. A. Hamilton.
    Vol. 1. English—German—French. 1s.
42. ————— Vol. 2. German—French—English. 1s.
43. ————— Vol. 3. French—German—English. 1s.
    ————— Complete in 1 vol. 3s.; cloth boards, 4s.
    —————————————, with Grammar. Cloth boards, 5s.
44. HEBREW DICTIONARY, by Bresslau. Vol. 1. Heb.—Eng. 6s.
    —————————————, with Grammar. 7s.
46. ————————— Vol. 2. English—Hebrew. 3s.
    ————— Complete, with Grammar, in 2 vols. Cloth boards, 12s.
46*. ————— GRAMMAR, by Dr. Bresslau. 1s.
47. FRENCH AND ENGLISH PHRASE BOOK. 1s.
48. COMPOSITION AND PUNCTUATION, by J. Brenan. 1s.
49. DERIVATIVE SPELLING BOOK, by J. Rowbotham. 1s. 6d.
50. DATES AND EVENTS. A Tabular View of English History,
    with Tabular Geography, by Edgar H. Rand. [In Preparation.
ART OF EXTEMPORE SPEAKING. Hints for the
    Pulpit, the Senate, and the Bar, by M. Bautain, Professor at
    the Sorbonne, &c.   [Just ready.

THE SCHOOL MANAGERS' SERIES
OF
READING BOOKS,
Adapted to the Requirements of the New Code of 1871.

Edited by the Rev. A. R. GRANT, Rector of Hitcham, and Honorary
    Canon of Ely; formerly H.M. Inspector of Schools.

	s.	d.		s.	d.
FIRST STANDARD	0	3	FOURTH STANDARD	0	10
SECOND „	0	6	FIFTH „	1	0
THIRD „	0	8	SIXTH „	1	2

PUBLISHED BY LOCKWOOD & CO.,

LATIN AND GREEK CLASSICS,

WITH EXPLANATORY NOTES IN ENGLISH.

LATIN SERIES.

1. A NEW LATIN DELECTUS, with Vocabularies and
Notes, by H. Young . . . . . . 1s.
2. CÆSAR. De Bello Gallico; Notes by H. Young . . 2s.
3. CORNELIUS NEPOS; Notes by H. Young . . . 1s.
4. VIRGIL. The Georgics, Bucolics, and Doubtful Poems;
Notes by W. Rushton, M.A., and H. Young . 1s. 6d.
5. VIRGIL. Æneid; Notes by H. Young . . . 2s.
6. HORACE. Odes, Epodes, and Carmen Seculare, by H. Young 1s.
7. HORACE. Satires and Epistles, by W. B. Smith, M.A. 1s. 6d.
8. SALLUST. Catiline and Jugurthine War; Notes by
W. M. Donne, B.A. . . . . . . 1s. 6d.
9. TERENCE. Andria and Heautontimorumenos; Notes by
the Rev. J. Davies, M.A. . . . . . 1s. 6d.
10. TERENCE. Adelphi, Hecyra, and Phormio; Notes by
the Rev. J. Davies, M.A. . . . . . . 2s.
11. TERENCE. Eunuchus, by Rev. J. Davies, M.A. . 1s. 6d.
Nos. 9, 10, and 11 in 1 vol. cloth boards, 6s.
12. CICERO. Oratio Pro Sexto Roscio Amerino. Edited,
with Notes, &c., by J. Davies, M.A. (*Now ready.*) . . 1s.
14. CICERO. De Amicitia, de Senectute, and Brutus; Notes
by the Rev. W. B. Smith, M.A. . . . . . 2s.
16. LIVY. Books i., ii., by H. Young . . . . 1s. 6d.
16*. LIVY. Books iii., iv., v., by H. Young . . . 1s. 6d.
17. LIVY. Books xxi., xxii., by W. B. Smith, M.A. . 1s. 6d.
19. CATULLUS, TIBULLUS, OVID, and PROPERTIUS,
Selections from, by W. Bodham Donne . . . . 2s.
20. SUETONIUS and the later Latin Writers, Selections from,
by W. Bodham Donne . . . . 2s.
21. THE SATIRES OF JUVENAL, by T. H. S. Escott, M.A.,
of Queen's College, Oxford . . . . . 1s. 6d.

GREEK SERIES.

WITH EXPLANATORY NOTES IN ENGLISH.

1. A NEW GREEK DELECTUS, by H. Young .   1s.
2. XENOPHON. Anabasis, i. ii. iii., by H. Young . . 1s.
3. XENOPHON. Anabasis, iv. v. vi. vii., by H. Young . 1s.
4. LUCIAN. Select Dialogues, by H. Young . . . 1s.
5. HOMER. Iliad, i. to vi., by T. H. L. Leary, D.C.L.   1s. 6d.
6. HOMER. Iliad, vii. to xii., by T. H. L. Leary, D.C.L. 1s. 6d.
7. HOMER. Iliad, xiii. to xviii., by T.H.L. Leary, D.C.L. 1s. 6d.
8. HOMER. Iliad, xix. to xxiv., by T. H. L. Leary, D.C.L. 1s. 6d.
9. HOMER. Odyssey, i. to vi., by T. H. L. Leary, D.C.L. 1s. 6d.
10. HOMER. Odyssey, vii. to xii., by T. H. L. Leary, D.C.L. 1s. 6d.
11. HOMER. Odyssey, xiii. to xviii., by T.H.L.Leary, D.C.L. 1s. 6d.
12. HOMER. Odyssey, xix. to xxiv.; and Hymns, by T. H. L.
    Leary, D.C.L. . . . . 2s.
13. PLATO. Apologia, Crito, and Phædo, by J. Davies, M.A. 2s.
14. HERODOTUS, Books i. ii., by T. H. L. Leary, D.C.L. 1s. 6d.
15. HERODOTUS, Books iii. iv., by T. H. L. Leary, D.C.L. 1s. 6d.
16. HERODOTUS, Books v. vi. vii., by T. H. L. Leary, D.C.L. 1s. 6d.
17. HERODOTUS, Books viii. ix., and Index, by T. H. L.
    Leary, D.C.L. . . . . 1s. 6d.
18. SOPHOCLES. Œdipus Tyrannus, by H. Young . . 1s.
20. SOPHOCLES. Antigone, by J. Milner, B.A. . . .· 2s.
23. EURIPIDES. Hecuba and Medea, by W. B. Smith, M.A. 1s. 6d.
26. EURIPIDES. Alcestis, by J. Milner, B.A. . . . 1s.
30. ÆSCHYLUS. Prometheus Vinctus, by J. Davies, M.A. . 1s.
32. ÆSCHYLUS. Septem contra Thebas, by J. Davies, M.A. 1s.
40. ARISTOPHANES. Acharnenses, by C. S. D. Townshend,
    M.A. . . . . 1s. 6d.
41. THUCYDIDES. Peloponnesian War. Book i., by H. Young 1s.
42. XENOPHON. Panegyric on Agesilaus, by Ll. F. W. Jewitt 1s. 6d.

www.ingramcontent.com/pod-product-compliance
Lightning Source LLC
Chambersburg PA
CBHW021803190326
41518CB00007B/429